Active Calculus - Multivariable
2017 edition

Steve Schlicker, Lead Author and Editor
Department of Mathematics
Grand Valley State University
schlicks@gvsu.edu
http://faculty.gvsu.edu/schlicks/

David Austin, Contributing Author
http://merganser.math.gvsu.edu/david/

Matt Boelkins, Contributing Author
http://faculty.gvsu.edu/boelkinm/

July 8, 2017

Contents

Preface	v

9 Multivariable and Vector Functions — 1

- 9.1 Functions of Several Variables and Three Dimensional Space 1
- 9.2 Vectors 19
- 9.3 The Dot Product 31
- 9.4 The Cross Product 40
- 9.5 Lines and Planes in Space 50
- 9.6 Vector-Valued Functions 60
- 9.7 Derivatives and Integrals of Vector-Valued Functions 66
- 9.8 Arc Length and Curvature 79

10 Derivatives of Multivariable Functions — 93

- 10.1 Limits 93
- 10.2 First-Order Partial Derivatives 105
- 10.3 Second-Order Partial Derivatives 117
- 10.4 Linearization: Tangent Planes and Differentials 127
- 10.5 The Chain Rule 139
- 10.6 Directional Derivatives and the Gradient 147
- 10.7 Optimization 159
- 10.8 Constrained Optimization:Lagrange Multipliers 175

11 Multiple Integrals — 181

- 11.1 Double Riemann Sums and Double Integrals over Rectangles 181
- 11.2 Iterated Integrals 192

- 11.3 Double Integrals over General Regions 198
- 11.4 Applications of Double Integrals 208
- 11.5 Double Integrals in Polar Coordinates 218
- 11.6 Surfaces Defined Parametrically and Surface Area 226
- 11.7 Triple Integrals 235
- 11.8 Triple Integrals in Cylindrical and Spherical Coordinates 246
- 11.9 Change of Variables 257

Preface

A free and open-source calculus

As we have noted in the single-variable edition of *Active Calculus*, several fundamental ideas in calculus are more than 2000 years old, and calculus as a formal subdiscipline of mathematics was first introduced and developed in the late 1600s. Mathematicians agree that the subject has been understood – rigorously and by experts – since the mid 1800s. The discipline is one of our great human intellectual achievements.

While each author of a calculus textbook certainly offers their own creative perspective on the subject, it is hardly the case that many of the ideas the author presents are new. Indeed, the mathematics community broadly agrees on what the main ideas of calculus are, as well as their justification and their importance; the core parts of nearly all calculus textbooks are very similar. As such, it is our opinion that in the 21st century – an age where the internet permits seamless and immediate transmission of information – no one should be required to purchase a calculus text to read, to use for a class, or to find a coherent collection of problems to solve. Calculus belongs to humankind, not any individual author or publishing company. Thus, a main purpose of this work is to present a new multivariable calculus text that is *free*. In addition, instructors who are looking for a multivariable calculus text should have the opportunity to download the source files and make modifications that they see fit; thus this text is *open-source*.

Any professor or student may use an electronic version of the text for no charge. A .pdf copy of the text may be obtained by download from the Active Calculus home page, or directly from

$$\text{http://gvsu.edu/s/Wb}.$$

Because the text is open-source, any instructor may acquire the full set of source files, by request via email to Steve Schlicker.

This work is licensed under the Creative Commons Attribution-NonCommercial-ShareAlike 3.0 Unported License. The graphic

that appears throughout the text shows that the work is licensed with the Creative Commons, that the work may be used for free by any party so long as attribution is given to the author(s), that the work and its derivatives are used in the spirit of "share and share alike," and that no party may sell this work or any of its derivatives for profit. Full details may be found by visiting

> http://creativecommons.org/licenses/by-nc-sa/3.0/

or sending a letter to Creative Commons, 444 Castro Street, Suite 900, Mountain View, California, 94041, USA.

Active Calculus - Multivariable: our goals

In *Active Calculus - Multivariable*, we endeavor to actively engage students in learning the subject through an activity-driven approach in which the vast majority of the examples are completed by students. Where many texts present a general theory of calculus followed by substantial collections of worked examples, we instead pose problems or situations, consider possibilities, and then ask students to investigate and explore. Following key activities or examples, the presentation normally includes some overall perspective and a brief synopsis of general trends or properties, followed by formal statements of rules or theorems. While we often offer plausibility arguments for such results, rarely do we include formal proofs. It is not the intent of this text for the instructor or author to *demonstrate* to students that the ideas of calculus are coherent and true, but rather for students to *encounter* these ideas in a supportive, leading manner that enables them to begin to understand for themselves why calculus is both coherent and true.

This approach is consistent with the following goals:

- To have students engage in an active, inquiry-driven approach, where learners strive to construct solutions and approaches to ideas on their own, with appropriate support through questions posed, hints, and guidance from the instructor and text.

- To build in students intuition for why the main ideas in multivariable calculus are natural and true. We strive to accomplish this by using specific cases to highlight the ideas for the general situation using contexts that are common and familiar.

- To challenge students to acquire deep, personal understanding of multivariable calculus through reading the text and completing preview activities on their own, through working on activities in small groups in class, and through doing substantial exercises outside of class time.

- To strengthen students' written and oral communicating skills by having them write about and explain aloud the key ideas of multivariable calculus.

Features of the Text

Similar to the presentation of the single-variable *Active Calculus*, instructors and students alike will find several consistent features in the presentation, including:

- **Motivating Questions.** At the start of each section, we list *motivating questions* that provide motivation for why the following material is of interest to us. One goal of each section is to answer each of the motivating questions.

- **Preview Activities.** Each section of the text begins with a short introduction, followed by a *preview activity*. This brief reading and the preview activity are designed to foreshadow the upcoming ideas in the remainder of the section; both the reading and preview activity are intended to be accessible to students *in advance* of class, and indeed to be completed by students before a day on which a particular section is to be considered.

- **Activities.** Every section in the text contains several *activities*. These are designed to engage students in an inquiry-based style that encourages them to construct solutions to key examples on their own, working either individually or in small groups.

- **Exercises.** There are dozens of calculus texts with (collectively) tens of thousands of exercises. Rather than repeat standard and routine exercises in this text, we recommend the use of WeBWorK with its access to the National Problem Library and its many multivariable calculus problems. In this text, there are a small number of challenging exercises in each section. Almost every such exercise has multiple parts, requires the student to connect several key ideas, and expects that the student will do at least a modest amount of writing to answer the questions and explain their findings. For instructors interested in a more conventional source of exercises, consider the freely available text by Gilbert Strang of MIT, available in .pdf format from the MIT open courseware site via `http://gvsu.edu/s/bh`.

- **Graphics.** As much as possible, we strive to demonstrate key fundamental ideas visually, and to encourage students to do the same. Throughout the text, we use full-color graphics to exemplify and magnify key ideas, and to use this graphical perspective alongside both numerical and algebraic representations of calculus. The figures and the software to generate them have been created by David Austin.

- **Links to Java Applets.** Many of the ideas of multivariable calculus are best understood dynamically, and there are a number of applets referenced in the text that can be used by instructors and students to assist in the investigations and demonstrations. The use of these freely available applets is in accord with our philosophy that no one should be required to purchase materials to learn calculus. We are indebted to everyone who allows their expertise to be openly shared.

- **Summary of Key Ideas.** Each section concludes with a summary of the key ideas encountered in the preceding section; this summary normally reflects responses to the motivating questions that began the section.

How to Use this Text

This text may be used as a stand-alone textbook for a standard multivariable calculus course or as a supplement to a more traditional text.

Electronically

Because students and instructors alike have access to the book in .pdf format, there are several advantages to the text over a traditional print text. One is that the text may be projected on a screen in the classroom (or even better, on a whiteboard) and the instructor may reference ideas in the text directly, add comments or notation or features to graphs, and indeed write right on the text itself. Students can do likewise, choosing to print only whatever portions of the text are needed for them. In addition, the electronic version of the text includes live html links to java applets, so student and instructor alike may follow those links to additional resources that lie outside the text itself. Finally, students can have access to a copy of the text anywhere they have an internet-enabled device.

Activities Workbook

Each section of the text has a preview activity and at least three in-class activities embedded in the discussion. As it is the expectation that students will complete all of these activities, it is ideal for them to have room to work on them adjacent to the problem statements themselves. As a separate document, we have compiled a workbook of activities that includes only the individual activity prompts, along with space provided for students to write their responses. This workbook is the one printing expense that students will almost certainly have to undertake, and is available upon request to the authors.

There are also options in the source files for compiling the activities workbook with hints for each activity, or even full solutions. These options can be engaged at the instructor's discretion.

Community of Users

Because this text is free and open-source, we hope that as people use the text, they will contribute corrections, suggestions, and new material. At this time, the best way to communicate typographical errors or suggestions for changes is by email to Matt Boelkins at `boelkinm@gvsu.edu`. At Matt's blog, `http://opencalculus.wordpress.com/`, we also share new developments, post feedback received by email, and include other points of discussion; readers may post additional comments and feedback.

Acknowledgments

This text is an extension of the single variable *Active Calculus* by Matt Boelkins. The initial drafts of this multivariable edition were written by Steve Schlicker; editing and revisions were made by David Austin and Matt Boelkins. David Austin is responsible for the beautiful full-color .eps graphics in the text. Many of our colleagues at GVSU have shared their ideas and resources, which undoubtedly had a significant influence on the product. We thank them for all of their support.

In advance, we also thank our colleagues throughout the mathematical community who will read, edit, and use this book, and hence contribute to its improvement through ongoing discussion. The following people have used early drafts of this text and have generously offered suggestions that have improved the text.

Jon Barker and students	St. Ignatius High School, Cleveland, OH
Brian Drake	Grand Valley State University
Brian Gleason	Nevada State College
Mitch Keller	Washington & Lee University

In addition, we offer thanks to Lars Jensen for giving us permission to use his photograph of the Lake of the Clouds in the Porcupine Mountains State Park in Michigan's upper peninsula for our cover. A contour plot of a portion of the Porcupine Mountains State Park is the focus of an activity in Section 9.1.

The authors take responsibility for all errors or inconsistencies in the text, and welcome reader and user feedback to correct them, along with other suggestions to improve the text.

David Austin, Matt Boelkins, Steven Schlicker, Allendale, MI, August, 2016.

x

Chapter 9

Multivariable and Vector Functions

9.1 Functions of Several Variables and Three Dimensional Space

Motivating Questions

In this section, we strive to understand the ideas generated by the following important questions:

- What is a function of several variables? What do we mean by the domain of a function of several variables?

- How do we find the distance between two points in $\mathbb{R}^3$? What is the equation of a sphere in $\mathbb{R}^3$?

- What is a trace of a function of two variables? What does a trace tell us about a function?

- What is a level curve of a function of two variables? What does a level curve tell us about a function?

Introduction

Throughout our mathematical careers we have studied functions of a single variable. We define a function of one variable as a rule that assigns exactly one output to each input. We analyze these functions by looking at their graphs, calculating limits, differentiating, integrating, and more. In this and following sections, we will study functions whose input is defined in terms of more than one variable, and then analyze these functions by looking at their graphs, calculating limits, differentiating, integrating, and more. We will see that many of the ideas from single variable calculus translate well to functions of several variables, but we will have to make some adjustments as well.

Preview Activity 9.1. When people buy a large ticket item like a car or a house, they often take out a loan to make the purchase. The loan is paid back in monthly installments until the entire amount of the loan, plus interest, is paid. The monthly payment that the borrower has to make depends on the amount P of money borrowed (called the principal), the duration t of the loan in years, and the interest rate r. For example, if we borrow \$18,000 to buy a car, the monthly payment M that we need to make to pay off the loan is given by the formula

$$M = \frac{1500r}{1 - \frac{1}{\left(1+\frac{r}{12}\right)^{12t}}}.$$

The variables r and t are independent of each other, so using functional notation we write

$$M(r,t) = \frac{1500r}{1 - \frac{1}{\left(1+\frac{r}{12}\right)^{12t}}}.$$

(a) Find the monthly payments on this loan if the interest rate is 6% and the duration of the loan is 5 years.

(b) Evaluate $M(0.05, 4)$. Explain in words what this calculation represents.

(c) Now consider only loans where the interest rate is 5%. Calculate the monthly payments as indicated in Table 9.1. Round payments to the nearest penny.

Duration (in years)	2	3	4	5	6
Monthly payments (dollars)					

Table 9.1: Monthly payments at an interest rate of 5%.

(d) Now consider only loans where the duration is 3 years. Calculate the monthly payments as indicated in Table 9.2. Round payments to the nearest penny.

Interest rate	0.03	0.05	0.07	0.09	0.11
Monthly payments (dollars)					

Table 9.2: Monthly payments over three years.

(e) Describe as best you can the combinations of interest rates and durations of loans that result in a monthly payment of \$200.

9.1. FUNCTIONS OF SEVERAL VARIABLES AND THREE DIMENSIONAL SPACE

Functions of Several Variables

Suppose we launch a projectile, using a golf club, a cannon, or some other device, from ground level. Under ideal conditions (ignoring wind resistance, spin, or any other forces except the force of gravity) the horizontal distance the object will travel depends on the initial velocity x the object is given, and the angle y at which it is launched. If we let f represent the horizontal distance the object travels (its range), then f is a function of the two variables x and y, and we represent f in functional notation by

$$f(x,y) = \frac{x^2 \sin(2y)}{g},$$

where g is the acceleration due to gravity.[1]

> **Definition 9.1.** A **function f of two independent variables** is a rule that assigns to each ordered pair (x, y) in some set D exactly one real number $f(x, y)$.

There is, of course, no reason to restrict ourselves to functions of only two variables—we can use any number of variables we like. For example,

$$f(x, y, z) = x^2 - 2xz + \cos(y)$$

defines f as a function of the three variables x, y, and z. In general, a function of n independent variables is a rule that assigns to an ordered n-tuple $(x_1, x_2, \ldots, x_n)$ in some set D exactly one real number.

As with functions of a single variable, it is important to understand the set of inputs for which the function is defined.

> **Definition 9.2.** The **domain** of a function f is the set of all inputs at which the function is defined.

Activity 9.1. REDO!

Identify the domain of each of the following functions. Draw a picture of each domain in the x-y plane.

(a) $f(x, y) = x^2 + y^2$

(b) $f(x, y) = \sqrt{x^2 + y^2}$

(c) $Q(x, y) = \frac{x+y}{x^2 - y^2}$

(d) $s(x, y) = \frac{1}{\sqrt{1 - xy^2}}$

◁

---

[1] We will derive this equation in a later section.

9.1. FUNCTIONS OF SEVERAL VARIABLES AND THREE DIMENSIONAL SPACE

Representing Functions of Two Variables

One of the techniques we use to study functions of one variable is to create a table of values. We can do the same for functions of two variables, except that our tables will have to allow us to keep track of both input variables. We can do this with a 2-dimensional table, where we list the x-values down the first column and the y-values across the first row. Let f be the function defined by $f(x,y) = \frac{x^2 \sin(2y)}{g}$ that gives the range of a projectile as a function of the initial velocity x and launch angle y of the projectile. The value $f(x,y)$ is then displayed in the location where the x row intersects the y column, as shown in Table 9.3 (where we measure x in feet and y in radians).

$x \longrightarrow$
$y \downarrow$

$f(0.2, 250) = (0.2)^2 \sin(2 \cdot 250)$

$x \backslash y$	0.2	0.4	0.6	0.8	1.0	1.2	1.4
25	7.6	14.0	18.2	19.5	17.8	13.2	6.5
50	30.4	56.0	72.8	78.1	71.0	52.8	26.2
75	68.4	126.1	163.8	175.7	159.8	118.7	58.9
100	121.7	224.2	291.3	312.4	284.2	211.1	104.7
125	190.1	350.3	455.1	488.1	444.0	329.8	163.6
150	273.8	504.4	655.3	702.8	639.3	474.9	235.5
175	372.7	686.5	892.0	956.6	870.2	646.4	320.6
200	486.8	896.7	1165.0	1249.5	1136.6	844.3	418.7
225	616.2	1134.9	1474.5	1581.4	1438.5	1068.6	530.0
250							

Table 9.3: Values of $f(x,y) = \frac{x^2 \sin(2y)}{g}$.

Activity 9.2.

Complete the last row in Table 9.3 to provide the needed values of the function f.

◁

If f is a function of a single variable x, then we define the graph of f to be the set of points of the form $(x, f(x))$, where x is in the domain of f. We then plot these points using the coordinate axes in order to visualize the graph. We can do a similar thing with functions of several variables. Table 9.3 identifies points of the form $(x, y, f(x,y))$, and we define the graph of f to be the set of these points.

> **Definition 9.3.** The **graph** of a function $f = f(x,y)$ is the set of points of the form $(x, y, f(x,y))$, where the point (x,y) is in the domain of f.

We also often refer to the graph of a function f of two variables as the *surface* generated by f. Points in the form $(x, y, f(x,y))$ are in three dimensions, so plotting these points takes a bit more work than graphs of functions in two dimensions. To plot these three-dimensional points, we need to set up a coordinate system with three mutually perpendicular axes – the x-axis, the

9.1. FUNCTIONS OF SEVERAL VARIABLES AND THREE DIMENSIONAL SPACE

y-axis, and the z-axis (called the *coordinate axes*). There are essentially two different ways we could set up a 3D coordinate system, as shown in Figures 9.1 and 9.2; thus, before we can proceed, we need to establish a convention.

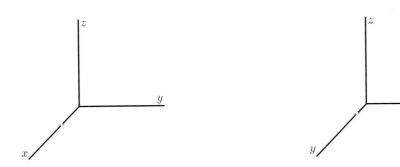

may's preference: ↓

Figure 9.1: A right hand system   Figure 9.2: A left hand system

The distinction between these two figures is subtle, but important. In the coordinate system shown in 9.1, imagine that you are sitting on the positive z-axis next to the label "z." Looking down at the x- and y-axes, you see that the y-axis is obtained by rotating the x-axis by 90° in the *counterclockwise* direction. Again sitting on the positive z-axis in Figure 9.2, you see that the y-axis is obtained by rotating the x-axis by 90° in the *clockwise* direction.

We call the coordinate system in 9.1 a *right-hand system*; if we point the index finger of our *right* hand along the positive x-axis and our middle finger along the positive y-axis, then our thumb points in the direction of the positive z-axis. Following mathematical conventions, we choose to use a right-hand system throughout this book.

Now that we have established a convention for a right hand system, we can draw a graph of the range function defined by $f(x, y) = \frac{x^2 \sin(2y)}{g}$. Note that the function f is continuous in both variables, so when we plot these points in the right hand coordinate system, we can connect them all to form a surface in 3-space. The graph of the range function f is shown in Figure 9.3.

There are many graphing tools available for drawing three-dimensional surfaces.[2] Since we will be able to visualize graphs of functions of two variables, but not functions of more than two variables, we will primarily deal with functions of two variables in this course. It is important to note, however, that the techniques we develop apply to functions of any number of variables.

Notation: We let $\mathbb{R}^2$ denote the set of all ordered pairs of real numbers in the plane (two copies of the real number system) and let $\mathbb{R}^3$ represent the set of all ordered triples of real numbers (which constitutes three-space).

---

[2]e.g., Wolfram Alpha and `http://web.monroecc.edu/manila/webfiles/calcNSF/JavaCode/CalcPlot3D.htm`

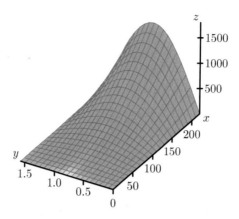

Figure 9.3: The range surface.

Some Standard Equations in Three-Space

In addition to graphing functions, we will also want to understand graphs of some simple equations in three dimensions. For example, in $\mathbb{R}^2$, the graphs of the equations $x = a$ and $y = b$, where a and b are constants, are lines parallel to the coordinate axes. In the next activity we consider their three-dimensional analogs.

Activity 9.3.

(a) Consider the set of points (x, y, z) that satisfy the equation $x = 2$. Describe this set as best you can.

(b) Consider the set of points (x, y, z) that satisfy the equation $y = -1$. Describe this set as best you can.

(c) Consider the set of points (x, y, z) that satisfy the equation $z = 0$. Describe this set as best you can.

◁

Activity 9.3 shows that the equations where one independent variable is constant lead to planes parallel to ones that result from a pair of the coordinate axes. When we make the constant 0, we get the *coordinate planes*. The xy-plane satisfies $z = 0$, the xz-plane satisfies $y = 0$, and the yz-plane satisfies $z = 0$ (see Figure 9.4).

On a related note, we define a circle in $\mathbb{R}^2$ as the set of all points equidistant from a fixed point. In $\mathbb{R}^3$, we call the set of all points equidistant from a fixed point a *sphere*. To find the equation of a sphere, we need to understand how to calculate the distance between two points in three-space, and we explore this idea in the next activity.

Activity 9.4.

Let $P = (x_0, y_0, z_0)$ and $Q = (x_1, y_1, z_1)$ be two points in $\mathbb{R}^3$. These two points form opposite vertices of a rectangular box whose sides are planes parallel to the coordinate planes as illus-

9.1. FUNCTIONS OF SEVERAL VARIABLES AND THREE DIMENSIONAL SPACE

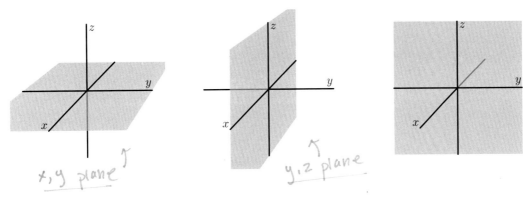

Figure 9.4: The coordinate planes.

trated in Figure 9.5, and the distance between P and Q is the length of the diagonal shown in Figure 9.5.

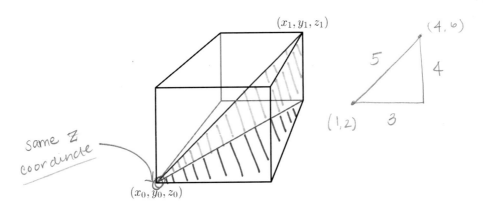

Figure 9.5: The distance formula in $\mathbb{R}^3$.

(a) Consider one of the right triangles in the base of the box whose hypotenuse is shown as the red line in Figure 9.5. What are the vertices of this triangle? Since this right triangle lies in a plane, we can use the Pythagorean Theorem to find a formula for the length of the hypotenuse of this triangle. Find such a formula, which will be in terms of x_0, y_0, x_1, and y_1.

(b) Now notice that the triangle whose hypotenuse is the blue segment connecting the points P and Q with a leg as the hypotenuse of the triangle found in part (a) lies entirely in a plane, so we can again use the Pythagorean Theorem to find the length of its hypotenuse. Explain why the length of this hypotenuse, which is the distance between the points P and Q, is

$$\sqrt{(x_1 - x_0)^2 + (y_1 - y_0)^2 + (z_1 - z_0)^2}.$$

The formula developed in Activity 9.4 is important to remember.

> The distance between points $P = (x_0, y_0, z_0)$ and $Q = (x_1, y_1, z_1)$ (denoted as $|PQ|$) in $\mathbb{R}^3$ is given by the formula
>
> $$|PQ| = \sqrt{(x_1 - x_0)^2 + (y_1 - y_0)^2 + (z_1 - z_0)^2}. \qquad (9.2)$$

Equation (9.2) can be used to derive the formula for a sphere centered at a point (x_0, y_0, z_0) with radius r. Since the distance from any point (x, y, z) on such a sphere to the point (x_0, y_0, z_0) is r, the point (x, y, z) will satisfy the equation

$$\sqrt{(x - x_0)^2 + (y - y_0)^2 + (z - z_0)^2} = r$$

Squaring both sides, we come to the standard equation for a sphere.

> The equation of a sphere with center (x_0, y_0, z_0) and radius r is
>
> $$(x - x_0)^2 + (y - y_0)^2 + (z - z_0)^2 = r^2.$$

This makes sense if we compare this equation to its two-dimensional analogue, the equation of a circle of radius r in the plane centered at (x_0, y_0):

$$(x - x_0)^2 + (y - y_0)^2 = r^2.$$

Traces

When we study functions of several variables we are often interested in how each individual variable affects the function in and of itself. In Preview Activity 9.1, we saw that the monthly payment on an $18,000 loan depends on the interest rate and the duration of the loan. However, if we fix the interest rate, the monthly payment depends only on the duration of the loan, and if we set the duration the payment depends only on the interest rate. This idea of keeping one variable constant while we allow the other to change will be an important tool for us when studying functions of several variables.

As another example, consider again the range function f defined by

$$f(x, y) = \frac{x^2 \sin(2y)}{g}$$

where x is the initial velocity of an object in feet per second, y is the launch angle in radians, and g is the acceleration due to gravity (32 feet per second squared). If we hold the launch angle constant at $y = 0.6$ radians, we can consider f a function of the initial velocity alone. In this case we have

$$f(x) = \frac{x^2}{32} \sin(2 \cdot 0.6).$$

9.1. FUNCTIONS OF SEVERAL VARIABLES AND THREE DIMENSIONAL SPACE

We can plot this curve on the surface by tracing out the points on the surface when $y = 0.6$, as shown in Figure 9.6. The graph and the formula clearly show that f is quadratic in the x-direction. More descriptively, as we increase the launch velocity while keeping the launch angle constant, the range increases proportional to the square of the initial velocity.

Similarly, if we fix the initial velocity at 150 feet per second, we can consider the range as a function of the launch angle only. In this case we have

$$f(y) = \frac{150^2 \sin(2y)}{32}.$$ parabola (x^k) · sin(x)

We can again plot this curve on the surface by tracing out the points on the surface when $x = 150$, as shown in Figure 9.7. The graph and the formula clearly show that f is sinusoidal in the y-direction. More descriptively, as we increase the launch angle while keeping the initial velocity constant, the range is proportional to the sine of twice the launch angle.

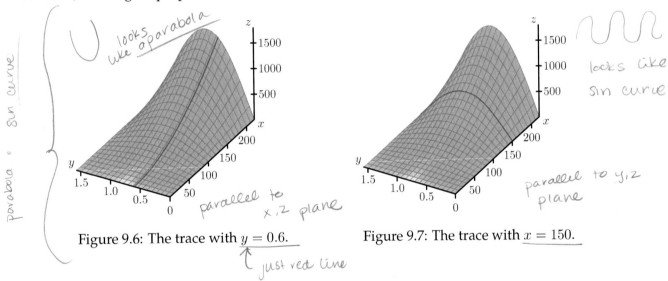

Figure 9.6: The trace with $y = 0.6$.   Figure 9.7: The trace with $x = 150$.

just red line

The curves we define when we fix one of the independent variables in our two variable function are called *traces*.

> **Definition 9.4.** A **trace** of a function f of two independent variables x and y is a curve of the form $z = f(c, y)$ or $z = f(x, c)$, where c is a constant.

Understanding trends in the behavior of functions of two variables can be challenging, as can sketching their graphs; traces help us with each of these tasks.

Activity 9.5.

In the following questions, we investigate the use of traces to better understand a function through both tables and graphs.

(a) Identify the $y = 0.6$ trace for the range function f defined by $f(x, y) = \frac{x^2 \sin(2y)}{g}$ by highlighting or circling the appropriate cells in Table 9.3. Write a sentence to describe the behavior of the function along this trace.

(b) Identify the $x = 150$ trace for the range function by highlighting or circling the appropriate cells in Table 9.3. Write a sentence to describe the behavior of the function along this trace.

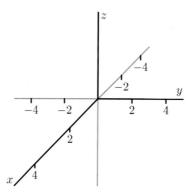

Figure 9.8: Coordinate axes to sketch traces.

(c) For the function g defined by $g(x,y) = x^2 + y^2 + 1$, explain the type of function that each trace in the x direction will be (keeping y constant). Plot the $y = -4$, $y = -2$, $y = 0$, $y = 2$, and $y = 4$ traces in 3-dimensional coordinate system provided in Figure 9.8.

(d) For the function g defined by $g(x,y) = x^2 + y^2 + 1$, explain the type of function that each trace in the y direction will be (keeping x constant). Plot the $x = -4$, $x = -2$, $x = 0$, $x - 2$, and $x = 4$ traces in 3-dimensional coordinate system in Figure 9.8.

(e) Describe the surface generated by the function g.

◁

Contour Maps and Level Curves

We have all seen topographic maps such as the one of the Porcupine Mountains in the upper peninsula of Michigan shown in Figure 9.9.[3] The curves on these maps show the regions of constant altitude. The contours also depict changes in altitude: contours that are close together signify steep ascents or descents, while contours that are far apart indicate only slight changes in elevation. Thus, contour maps tell us a lot about three-dimensional surfaces. Mathematically, if $f(x,y)$ represents the altitude at the point (x,y), then each contour is the graph of an equation of the form $f(x,y) = k$, for some constant k.

---

[3]Map source: Michigan Department of Natural Resources, https://www.michigan.gov/dnr/0,4570, 7-153-10369_46675_58093---,00.html, with permission of the Michigan DNR and Bob Wild. A picture of Lake of the Clouds (fro Lars Jensen) in shown on the cover.

9.1. FUNCTIONS OF SEVERAL VARIABLES AND THREE DIMENSIONAL SPACE

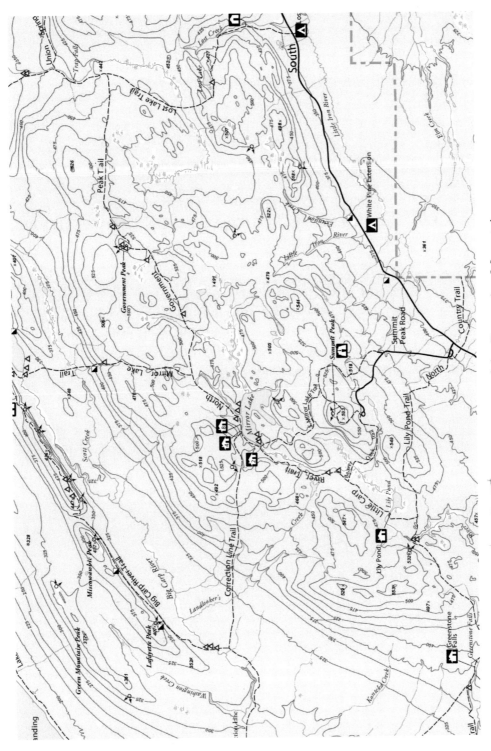

Figure 9.9: Contour map of the Porcupine Mountains.

Activity 9.6.

On the topographical map of the Porcupine Mountains in Figure 9.9,

(a) identify the highest and lowest points you can find;

(b) from a point of your choice, determine a path of steepest ascent that leads to the highest point;

(c) from that same initial point, determine the least steep path that leads to the highest point.

◁

Definition 9.5. A **level curve (or contour)** of a function f of two independent variables x and y is a curve of the form $k = f(x, y)$, where k is a constant.

Topographical maps can be used to create a three-dimensional surface from the two-dimensional contours or level curves. For example, level curves of the range function defined by $f(x, y) = \frac{x^2 \sin(2y)}{32}$ plotted in the xy-plane are shown in Figure 9.10. If we lift these contours and plot them at their respective heights, then we get a picture of the surface itself, as illustrated in Figure 9.11.

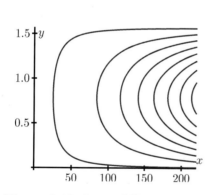

Figure 9.10: Several level curves.

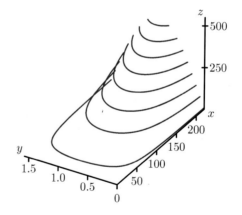

Figure 9.11: Level curves at the appropriate height.

The use of level curves and traces can help us construct the graph of a function of two variables.

Activity 9.7.

(a) Let $f(x, y) = x^2 + y^2$. Draw the level curves $f(x, y) = k$ for $k = 1$, $k = 2$, $k = 3$, and $k = 4$ on the left set of axes given in Figure 9.12. (You decide on the scale of the axes.) Explain what the surface defined by f looks like.

9.1. FUNCTIONS OF SEVERAL VARIABLES AND THREE DIMENSIONAL SPACE

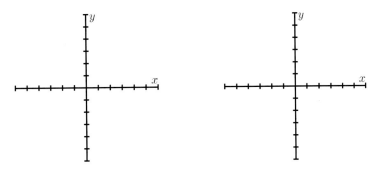

Figure 9.12: Left: Level curves for $f(x,y) = x^2 + y^2$. Right: Level curves for $g(x,y) = \sqrt{x^2 + y^2}$.

(b) Let $g(x,y) = \sqrt{x^2 + y^2}$. Draw the level curves $g(x,y) = k$ for $k = 1$, $k = 2$, $k = 3$, and $k = 4$ on the right set of axes given in Figure 9.12. (You decide on the scale of the axes.) Explain what the surface defined by g looks like.

(c) Compare and contrast the graphs of f and g. How are they alike? How are they different? Use traces for each function to help answer these questions.

◁

The traces and level curves of a function of two variables are curves in space. In order to understand these traces and level curves better, we will first spend some time learning about vectors and vector-valued functions in the next few sections and return to our study of functions of several variables once we have those more mathematical tools to support their study.

A gallery of functions HW describe what you see; why?

We end this section by considering a collection of functions and illustrating their graphs and some level curves.

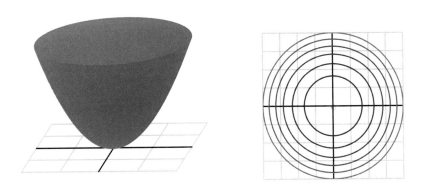

Figure 9.13: $z = x^2 + y^2$

$(0,3) = 9$
vs
$(0,0) = 0$

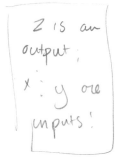

z is an output;
x, y are inputs!

 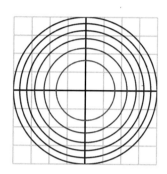

Figure 9.14: $z = 4 - (x^2 + y^2)$

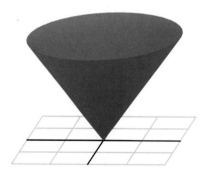 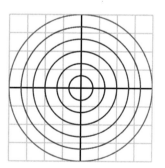

Figure 9.15: $z = \sqrt{x^2 + y^2}$

9.1. FUNCTIONS OF SEVERAL VARIABLES AND THREE DIMENSIONAL SPACE

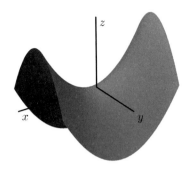 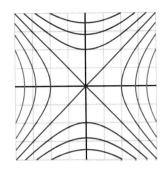

Figure 9.16: $z = x^2 - y^2$

 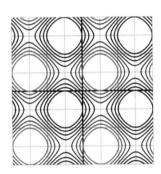

Figure 9.17: $z = \sin(x) + \sin(y)$

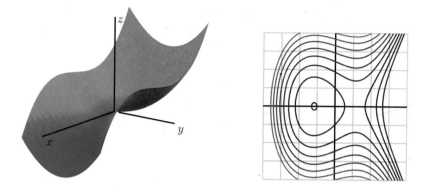

Figure 9.18: $z = y^2 - x^3 + x$

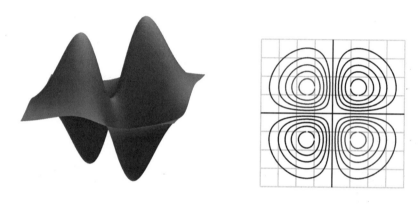

Figure 9.19: $z = xye^{-x^2-y^2}$

Summary

- A function f of several variables is a rule that assigns a unique number to an ordered collection of independent inputs. The domain of a function of several variables is the set of all inputs for which the function is defined.

- In $\mathbb{R}^3$, the distance between points $P = (x_0, y_0, z_0)$ and $Q = (x_1, y_1, z_1)$ (denoted as $|PQ|$) is given by the formula
$$|PQ| = \sqrt{(x_1 - x_0)^2 + (y_1 - y_0)^2 + (z_1 - z_0)^2}.$$
and thus the equation of a sphere with center (x_0, y_0, z_0) and radius r is
$$(x - x_0)^2 + (y - y_0)^2 + (z - z_0)^2 = r^2.$$

- A trace of a function f of two independent variables x and y is a curve of the form $z = f(c, y)$ or $z = f(x, c)$, where c is a constant. A trace tells us how the function depends on a single independent variable if we treat the other independent variable as a constant.

9.1. FUNCTIONS OF SEVERAL VARIABLES AND THREE DIMENSIONAL SPACE

- A level curve of a function f of two independent variables x and y is a curve of the form $k = f(x,y)$, where k is a constant. A level curve describes the set of inputs that lead to a specific output of the function.

Exercises

1. Find the equation of each of the following geometric objects.

    (a) The plane parallel to the x-y plane that passes through the point $(-4, 5, -12)$.

    (b) The plane parallel to the y-z plane that passes through the point $(7, -2, -3)$.

    (c) The sphere centered at the point $(2, 1, 3)$ and has the point $(-1, 0, -1)$ on its surface.

    (d) The sphere whose diameter has endpoints $(-3, 1, -5)$ and $(7, 9, -1)$.

2. The Ideal Gas Law, $PV = RT$, relates the pressure (P, in pascals), temperature (T, in Kelvin), and volume (V, in cubic meters) of 1 mole of a gas ($R = 8.314 \frac{\text{J}}{\text{mol K}}$ is the universal gas constant), and describes the behavior of gases that do not liquefy easily, such as oxygen and hydrogen. We can solve the ideal gas law for the volume and hence treat the volume as a function of the pressure and temperature:
$$V(P,T) = \frac{8.314T}{P}.$$

    (a) Explain in detail what the trace of V with $P = 1000$ tells us about a key relationship between two quantities.

    (b) Explain in detail what the trace of V with $T = 5$ tells us.

    (c) Explain in detail what the level curve $V = 0.5$ tells us.

    (d) Use 2 or three additional traces in each direction to make a rough sketch of the surface over the domain of V where P and T are each nonnegative. Write at least one sentence that describes the way the surface looks.

    (e) Based on all your work above, write a couple of sentences that describe the effects that temperature and pressure have on volume.

3. Consider the function h defined by $h(x,y) = 8 - \sqrt{4 - x^2 - y^2}$.

    (a) What is the domain of h? (Hint: describe a set of ordered pairs in the plane by explaining their relationship relative to a key circle.)

    (b) The *range* of a function is the set of all outputs the function generates. Given that the range of the square root function $g(t) = \sqrt{t}$ is the set of all nonnegative real numbers, what do you think is the range of h? Why?

    (c) Choose 4 different values from the range of h and plot the corresponding level curves in the plane. What is the shape of a typical level curve?

(d) Choose 5 different values of x (including at least one negative value and zero), and sketch the corresponding traces of the function h.

(e) Choose 5 different values of y (including at least one negative value and zero), and sketch the corresponding traces of the function h.

(f) Sketch an overall picture of the surface generated by h and write at least one sentence to describe how the surface appears visually. Does the surface remind you of a familiar physical structure in nature?

9.2 Vectors

Motivating Questions

In this section, we strive to understand the ideas generated by the following important questions:

- What is a vector?
- What does it mean for two vectors to be equal?
- How do we add two vectors together and multiply a vector by a scalar?
- How do we determine the magnitude of a vector? What is a unit vector, and how do we find a unit vector in the direction of a given vector?

Introduction

If we are at a point x in the domain of a function of one variable, there are only two directions in which we can move: in the positive or negative x-direction. If, however, we are at a point (x, y) in the domain of a function of two variables, there are many directions in which we can move. Thus, it is important for us to have a means to indicate direction, and we will do so using vectors.

Preview Activity 9.2. After working out, Sarah and John leave the Recreation Center on the Grand Valley State University Allendale campus (a map of which is given in Figure 9.20) to go to their next classes.[4] Suppose we record Sarah's movement on the map in a pair $\langle x, y \rangle$ (we will call this pair a *vector*), where x is the horizontal distance (in feet) she moves (with east as the positive direction) and y as the vertical distance (in feet) she moves (with north as the positive direction). We do the same for John. Throughout, use the legend to estimate your responses as best you can.

(a) What is the vector $\mathbf{v}_1 = \langle x, y \rangle$ that describes Sarah's movement if she walks directly in a straight line path from the Recreation Center to the entrance at the northwest end of Mackinac Hall? (Assume a straight line path, even if there are buildings in the way.) Explain how you found this vector. What is the total distance in feet between the Recreation Center and the entrance to Mackinac Hall? Measure the number of feet directly and then explain how to calculate this distance in terms of x and y.

(b) What is the vector $\mathbf{v}_2 = \langle x, y \rangle$ that describes John's change in position if he walks directly from the Recreation Center to Au Sable Hall? How many feet are there between Recreation Center to Au Sable Hall in terms of x and y?

(c) What is the vector $\mathbf{v}_3 = \langle x, y \rangle$ that describes the change in position if John walks directly from Au Sable Hall to the northwest entrance of Mackinac Hall to meet up with Sarah after

---

[4]GVSU campus map from http://www.gvsu.edu/homepage/files/pdf/maps/allendale.pdf, used with permission from GVSU, credit to illustrator Chris Bessert.

class? What relationship do you see among the vectors $\mathbf{v}_1$, $\mathbf{v}_2$, and $\mathbf{v}_3$? Explain why this relationship should hold.

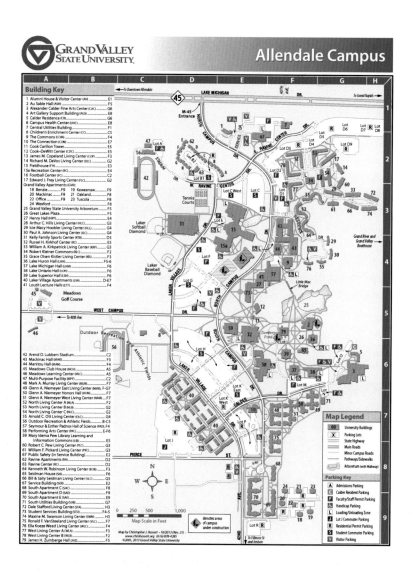

Figure 9.20: Grand Valley State University Allendale campus map.

9.2. VECTORS

Representations of Vectors

Preview Activity 9.2 shows how we can record the magnitude and direction of a change in position using an ordered pair of numbers $\langle x, y \rangle$. There are many other quantities, such as force and velocity, that possess the attributes of magnitude and direction, and we will call each such quantity a *vector*.

> **Definition 9.6.** A **vector** is any quantity that possesses the attributes of magnitude and direction.
> → which way?   how fast ←

We can represent a vector geometrically as a directed line segment, with the magnitude as the length of the segment and an arrowhead indicating direction, as shown in Figure 9.21.

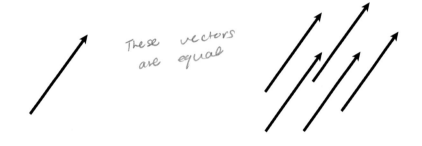

Figure 9.21: A vector.   Figure 9.22: Representations of the same vector.

According to the definition, a vector possesses the attributes of length (magnitude) and direction; the vector's position, however, is not mentioned. Consequently, we regard as equal any two vectors having the same magnitude and direction, as shown in Figure 9.22.

> Two vectors are equal provided they have the same magnitude and direction.

This means that the same vector may be drawn in the plane in many different ways. For instance, suppose that we would like to draw the vector $\langle 3, 4 \rangle$, which represents a horizontal change of three units and a vertical change of four units. We may place the *tail* of the vector (the point from which the vector originates) at the origin and the *tip* (the terminal point of the vector) at $(3, 4)$, as illustrated in Figure 9.23. A vector with its tail at the origin is said to be in *standard position*.

Alternatively, we may place the tail of the vector $\langle 3, 4 \rangle$ at another point, such as $Q(1, 1)$. After a displacement of three units to the right and four units up, the tip of the vector is at the point $R(4, 5)$ (see Figure 9.24).

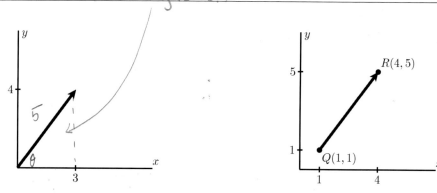

Figure 9.23: A vector in standard position   Figure 9.24: A vector between two points

In this example, the vector led to the directed line segment from Q to R, which we denote as $\overrightarrow{QR}$. We may also turn the situation around: given the two points Q and R, we obtain the vector $\langle 3, 4 \rangle$ because we move horizontally three units and vertically four units to get from Q to R. In other words, $\overrightarrow{QR} = \langle 3, 4 \rangle$. In general, the vector $\overrightarrow{QR}$ from the point $Q = (q_1, q_2)$ to $R = (r_1, r_2)$ is found by taking the difference of coordinates, so that

$$\overrightarrow{QR} = \langle r_1 - q_1, r_2 - q_2 \rangle.$$

We will use boldface letters to represent vectors, such as $\mathbf{v} = \langle 3, 4 \rangle$, to distinguish them from scalars. The entries of a vector are called its *components*: in the vector $\langle 3, 4 \rangle$, the x component is 3 and the y component is 4. We use pointed brackets $\langle , \rangle$ and the term *components* to distinguish a vector from a point $(,)$ and its *coordinates*. There is, however, a close connection between vectors and points. Given a point P, we will frequently consider the vector $\overrightarrow{OP}$ from the origin O to P. For instance, if $P = (3, 4)$, then $\overrightarrow{OP} = \langle 3, 4 \rangle$ as in Figure 9.25. In this way, we think of a point P as defining a vector $\overrightarrow{OP}$ whose components agree with the coordinates of P.

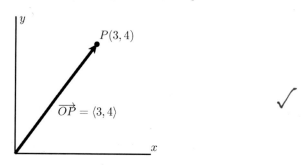

Figure 9.25: A point defines a vector

While we often illustrate vectors in the plane since it is easier to draw pictures, different situations call for the use of vectors in three or more dimensions. For instance, a vector $\mathbf{v}$ in n-

9.2. VECTORS

dimensional space, $\mathbb{R}^n$, has n components and may be represented as

$$\mathbf{v} = \langle v_1, v_2, v_3, \ldots, v_n \rangle.$$

⟩ = vector brackets!

The next activity will help us to become accustomed to vectors and operations on vectors in three dimensions.

Activity 9.8.

As a class, determine a coordinatization of your classroom, agreeing on some convenient set of axes (e.g., an intersection of walls and floor) and some units in the x, y, and z directions (e.g., using lengths of sides of floor, ceiling, or wall tiles). Let O be the origin of your coordinate system. Then, choose three points, A, B, and C in the room, and complete the following.

(a) Determine the coordinates of the points A, B, and C.

(b) Determine the components of the indicated vectors.

(i) $\overrightarrow{OA}$ (ii) $\overrightarrow{OB}$ (iii) $\overrightarrow{OC}$ (iv) $\overrightarrow{AB}$ (v) $\overrightarrow{AC}$ (vi) $\overrightarrow{BC}$

◁

Equality of Vectors

Because location is not mentioned in the definition of a vector, any two vectors that have the same magnitude and direction are equal. It is helpful to have an algebraic way to determine when this occurs. That is, if we know the components of two vectors $\mathbf{u}$ and $\mathbf{v}$, we will want to be able to determine algebraically when $\mathbf{u}$ and $\mathbf{v}$ are equal. There is an obvious set of conditions that we use.

Two vectors $\mathbf{u} = \langle u_1, u_2 \rangle$ and $\mathbf{v} = \langle v_1, v_2 \rangle$ in $\mathbb{R}^2$ are equal if and only if their corresponding components are equal: $u_1 = v_1$ and $u_2 = v_2$. More generally, two vectors $\mathbf{u} = \langle u_1, u_2, \ldots, u_n \rangle$ and $\mathbf{v} = \langle v_1, v_2, \ldots, v_n \rangle$ in $\mathbb{R}^n$ are equal if and only if $u_i = v_i$ for each possible value of i.

Operations on Vectors

Vectors are not numbers, but we can now represent them with components that are real numbers. As such, we naturally wonder if it is possible to add two vectors together, multiply two vectors, or combine vectors in any other ways. In this section, we will study two operations on vectors: vector addition and scalar multiplication. To begin, we investigate a natural way to add two vectors together, as well as to multiply a vector by a scalar.

Activity 9.9.

Let $\mathbf{u} = \langle 2, 3 \rangle$, $\mathbf{v} = \langle -1, 4 \rangle$.

(a) Using the two specific vectors above, what is the natural way to define the vector sum $\mathbf{u} + \mathbf{v}$?

$(x+x), (y+y) =$
$(2+(-1)), (3+4) =$ $\}$ $(-1, 7)$

(b) In general, how do you think the vector sum **a**+**b** of vectors $\mathbf{a} = \langle a_1, a_2 \rangle$ and $\mathbf{b} = \langle b_1, b_2 \rangle$ in $\mathbb{R}^2$ should be defined? Write a formal definition of a vector sum based on your intuition.

(c) In general, how do you think the vector sum **a** + **b** of vectors $\mathbf{a} = \langle a_1, a_2, a_3 \rangle$ and $\mathbf{b} = \langle b_1, b_2, b_3 \rangle$ in $\mathbb{R}^3$ should be defined? Write a formal definition of a vector sum based on your intuition.

(d) Returning to the specific vector $\mathbf{v} = \langle -1, 4 \rangle$ given above, what is the natural way to define the scalar multiple $\frac{1}{2}\mathbf{v}$? ↑ scalar $(-½, 2)$

(e) In general, how do you think a scalar multiple of a vector $\mathbf{a} = \langle a_1, a_2 \rangle$ in $\mathbb{R}^2$ by a scalar c should be defined? how about for a scalar multiple of a vector $\mathbf{a} = \langle a_1, a_2, a_3 \rangle$ in $\mathbb{R}^3$ by a scalar c? Write a formal definition of a scalar multiple of a vector based on your intuition.

◁

We can now add vectors and multiply vectors by scalars, and thus we can add together scalar multiples of vectors. This allows us to define *vector subtraction*, $\mathbf{v} - \mathbf{u}$, as the sum of $\mathbf{v}$ and -1 times $\mathbf{u}$, so that

$$\mathbf{v} - \mathbf{u} = \mathbf{v} + (-1)\mathbf{u}.$$

Using vector addition and scalar multiplication, we will often represent vectors in terms of the special vectors $\mathbf{i} = \langle 1, 0 \rangle$ and $\mathbf{j} = \langle 0, 1 \rangle$. For instance, we can write the vector $\langle a, b \rangle$ in $\mathbb{R}^2$ as

$$\langle a, b \rangle = a\langle 1, 0 \rangle + b\langle 0, 1 \rangle = a\mathbf{i} + b\mathbf{j},$$

which means that

$$\langle 2, -3 \rangle = 2\mathbf{i} - 3\mathbf{j}.$$

In the context of $\mathbb{R}^3$, we let $\mathbf{i} = \langle 1, 0, 0 \rangle$, $\mathbf{j} = \langle 0, 1, 0 \rangle$, and $\mathbf{k} = \langle 0, 0, 1 \rangle$, and we can write the vector $\langle a, b, c \rangle$ in $\mathbb{R}^3$ as

$$\langle a, b, c \rangle = a\langle 1, 0, 0 \rangle + b\langle 0, 1, 0 \rangle + c\langle 0, 0, 1 \rangle = a\mathbf{i} + b\mathbf{j} + c\mathbf{k}.$$

The vectors **i**, **j**, and **k** are called the *standard unit vectors*[5], and are important in the physical sciences.

Properties of Vector Operations

We know that the scalar sum $1 + 2$ is equal to the scalar sum $2 + 1$. This is called the *commutative* property of scalar addition. Any time we define operations on objects (like addition of vectors) we usually want to know what kinds of properties the operations have. For example, is addition of vectors a commutative operation? To answer this question we take two *arbitrary* vectors **v** and **u**

---

[5]As we will learn momentarily, unit vectors have length 1.

9.2. VECTORS

and add them together and see what happens. Let $\mathbf{v} = \langle v_1, v_2 \rangle$ and $\mathbf{u} = \langle u_1, u_2 \rangle$. Now we use the fact that $v_1, v_2, u_1,$ and u_2 are scalars, and that the addition of scalars is commutative to see that

$$\mathbf{v} + \mathbf{u} = \langle v_1, v_2 \rangle + \langle u_1, u_2 \rangle = \langle v_1 + u_1, v_2 + u_2 \rangle = \langle u_1 + v_1, u_2 + v_2 \rangle = \langle u_1, u_2 \rangle + \langle v_1, v_2 \rangle = \mathbf{u} + \mathbf{v}.$$

So the vector sum is a commutative operation. Similar arguments can be used to show the following properties of vector addition and scalar multiplication.

Let $\mathbf{v}$, $\mathbf{u}$, and $\mathbf{w}$ be vectors in $\mathbb{R}^n$ and let a and b be scalars. Then

1. $\mathbf{v} + \mathbf{u} = \mathbf{u} + \mathbf{v}$ ✓ *order doesn't matter*
2. $(\mathbf{v} + \mathbf{u}) + \mathbf{w} = \mathbf{v} + (\mathbf{u} + \mathbf{w})$ ✓
3. The vector $\mathbf{0} = \langle 0, 0, \ldots, 0 \rangle$ has the property that $\mathbf{v} + \mathbf{0} = \mathbf{v}$. The vector $\mathbf{0}$ is called the **zero vector**.
4. $(-1)\mathbf{v} + \mathbf{v} = \mathbf{0}$. The vector $(-1)\mathbf{v} = -\mathbf{v}$ is called the **additive inverse** of the vector $\mathbf{v}$.
5. $(a + b)\mathbf{v} = a\mathbf{v} + b\mathbf{v}$ *factoring*
6. $a(\mathbf{v} + \mathbf{u}) = a\mathbf{v} + a\mathbf{u}$ *factoring*
7. $(ab)\mathbf{v} = a(b\mathbf{v})$ ✓
8. $1\mathbf{v} = \mathbf{v}$. ✓

We verified the first property for vectors in $\mathbb{R}^2$; it is straightforward to verify that the rest of the eight properties just noted hold for all vectors in $\mathbb{R}^n$.

Geometric Interpretation of Vector Operations

Next, we explore a geometric interpretation of vector addition and scalar multiplication that allows us to visualize these operations. Let $\mathbf{u} = \langle 4, 6 \rangle$ and $\mathbf{v} = \langle 3, -2 \rangle$. Then $\mathbf{w} = \mathbf{u} + \mathbf{v} = \langle 7, 4 \rangle$, as shown on the left in Figure 9.26. $[(x+x), (y+y)]$

If we think of these vectors as displacements in the plane, we find a geometric way to envision vector addition. For instance, the vector $\mathbf{u} + \mathbf{v}$ will represent the displacement obtained by following the displacement $\mathbf{u}$ with the displacement $\mathbf{v}$. We may picture this by placing the tail of $\mathbf{v}$ at the tip of $\mathbf{u}$, as seen in the center of Figure 9.26.

Of course, vector addition is commutative so we obtain the same sum if we place the tail of $\mathbf{u}$ at the tip of $\mathbf{v}$. We therefore see that $\mathbf{u} + \mathbf{v}$ appears as the diagonal of the parallelogram determined by $\mathbf{u}$ and $\mathbf{v}$, as shown on the right of Figure 9.26.

Vector subtraction has a similar interpretation. On the left in Figure 9.27, we see vectors $\mathbf{u}$, $\mathbf{v}$, and $\mathbf{w} = \mathbf{u} + \mathbf{v}$. If we rewrite $\mathbf{v} = \mathbf{w} - \mathbf{u}$, we have the arrangement of Figure 9.28. In other words, to form the difference $\mathbf{w} - \mathbf{u}$, we draw a vector from the tip of $\mathbf{u}$ to the tip of $\mathbf{w}$.

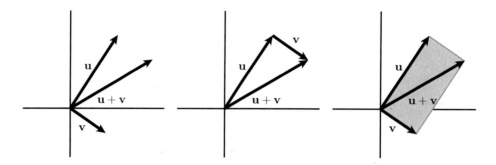

Figure 9.26: A vector sum (left), summing displacements (center), the parallelogram law (right)

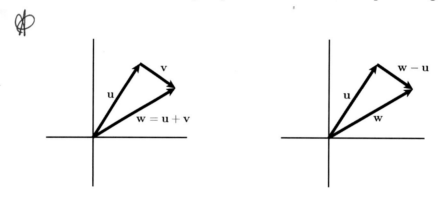

Figure 9.27: Vector addition

Figure 9.28: Vector subtraction

In a similar way, we may geometrically represent a scalar multiple of a vector. For instance, if $\mathbf{v} = \langle 2, 3 \rangle$, then $2\mathbf{v} = \langle 4, 6 \rangle$. As shown in Figure 9.29, multiplying $\mathbf{v}$ by 2 leaves the direction unchanged, but stretches $\mathbf{v}$ by 2. Also, $-2\mathbf{v} = \langle -4, -6 \rangle$, which shows that multiplying by a negative scalar gives a vector pointing in the opposite direction of $\mathbf{v}$.

9.2. VECTORS

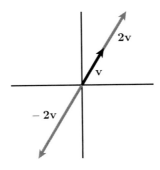

Figure 9.29: Scalar multiplication

Activity 9.10.

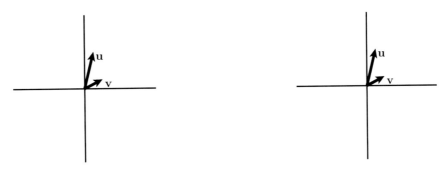

Figure 9.30   Figure 9.31

Suppose that **u** and **v** are the vectors shown in Figure 9.30.

(a) On Figure 9.30, sketch the vectors $\mathbf{u}+\mathbf{v}$, $\mathbf{v}-\mathbf{u}$, $2\mathbf{u}$, $-2\mathbf{u}$, and $-3\mathbf{v}$.

(b) What is $0\mathbf{v}$?

(c) On Figure 9.31, sketch the vectors $-3\mathbf{v}$, $-2\mathbf{v}$, $-1\mathbf{v}$, $2\mathbf{v}$, and $3\mathbf{v}$.

(d) Give a geometric description of the set of vectors $t\mathbf{v}$ where t is any scalar.

(e) On Figure 9.31, sketch the vectors $\mathbf{u}-3\mathbf{v}$, $\mathbf{u}-2\mathbf{v}$, $\mathbf{u}-\mathbf{v}$, $\mathbf{u}+\mathbf{v}$, and $\mathbf{u}+2\mathbf{v}$.

(f) Give a geometric description of the set of vectors $\mathbf{u}+t\mathbf{v}$ where t is any scalar.

◁

Done in class!!

The Magnitude of a Vector

By definition, vectors have both direction and magnitude (or length). We now investigate how to calculate the magnitude of a vector. Since a vector **v** can be represented by a directed line segment, we can use the distance formula to calculate the length of the segment. This length is the *magnitude* of the vector **v** and is denoted $|\mathbf{v}|$.

Activity 9.11. ← magnitude can't be negative

Figure 9.32: The vector defined by A and B.   Figure 9.33: An arbitrary vector, **v**.

(a) Let $A = (2, 3)$ and $B = (4, 7)$, as shown in Figure 9.32. Compute $|\overrightarrow{AB}|$.

(b) Let $\mathbf{v} = \langle v_1, v_2 \rangle$ be the vector in $\mathbb{R}^2$ with components v_1 and v_2 as shown in Figure 9.33. Use the distance formula to find a general formula for $|\mathbf{v}|$.

(c) Let $\mathbf{v} = \langle v_1, v_2, v_3 \rangle$ be a vector in $\mathbb{R}^3$. Use the distance formula to find a general formula for $|\mathbf{v}|$.

(d) Suppose that $\mathbf{u} = \langle 2, 3 \rangle$ and $\mathbf{v} = \langle -1, 2 \rangle$. Find $|\mathbf{u}|$, $|\mathbf{v}|$, and $|\mathbf{u} + \mathbf{v}|$. Is it true that $|\mathbf{u} + \mathbf{v}| = |\mathbf{u}| + |\mathbf{v}|$?

(e) Under what conditions will $|\mathbf{u} + \mathbf{v}| = |\mathbf{u}| + |\mathbf{v}|$? (Hint: Think about how **u**, **v**, and **u** + **v** form the sides of a triangle.) magnitudes don't need to be same but direction does!

(f) With the vector $\mathbf{u} = \langle 2, 3 \rangle$, find the lengths of $2\mathbf{u}$, $3\mathbf{u}$, and $-2\mathbf{u}$, respectively, and use proper notation to label your results.

(g) If t is any scalar, how is $|t\mathbf{u}|$ related to $|\mathbf{u}|$?

(h) A **unit vector** is a vector whose magnitude is 1. Of the vectors **i**, **j**, and **i** + **j**, which are unit vectors?

(i) Find a unit vector **v** whose direction is the same as $\mathbf{u} = \langle 2, 3 \rangle$. (Hint: Consider the result of part (g).)

Summary

- A vector is any object that possesses the attributes of magnitude and direction. Examples of vector quantities are position, velocity, acceleration, and force.

9.2. VECTORS

- Two vectors are equal if they have the same direction and magnitude. Notice that position is not considered, so a vector is independent of its location.

- If $\mathbf{u} = \langle u_1, u_2, \ldots, u_n \rangle$ and $\mathbf{v} = \langle v_1, v_2, \ldots, v_n \rangle$ are two vectors in $\mathbb{R}^n$, then their vector sum is the vector
$$\mathbf{u} + \mathbf{v} = \langle u_1 + v_1, u_2 + v_2, \ldots, u_n + v_n \rangle.$$

If $\mathbf{u} = \langle u_1, u_2, \ldots, u_n \rangle$ is a vector in $\mathbb{R}^n$ and c is a scalar, then the scalar multiple $c\mathbf{u}$ is the vector
$$c\mathbf{u} = \langle cu_1, cu_2, \ldots, cu_n \rangle.$$

- The magnitude of the vector $\mathbf{v} = \langle v_1, v_2, \ldots, v_n \rangle$ in $\mathbb{R}^n$ is the scalar
$$|\mathbf{v}| = \sqrt{v_1^2 + v_2^2 + \cdots + v_n^2}.$$

A vector $\mathbf{u}$ is a unit vector provided that $|\mathbf{u}| = 1$. If $\mathbf{v}$ is a nonzero vector, then the vector $\frac{\mathbf{v}}{|\mathbf{v}|}$ is a unit vector with the same direction as $\mathbf{v}$. *divide by its magntiutde to mak a vector a unit vector!*

Exercises

1. Let $\mathbf{v} = \langle 1, -2 \rangle$, $\mathbf{u} = \langle 0, 4 \rangle$, and $\mathbf{w} = \langle -5, 7 \rangle$.

    (a) Determine the components of the vector $\mathbf{u} - \mathbf{v}$.

    (b) Determine the components of the vector $2\mathbf{v} - 3\mathbf{u}$.

    (c) Determine the components of the vector $\mathbf{v} + 2\mathbf{u} - 7\mathbf{w}$.

    (d) Determine scalars a and b such that $a\mathbf{v} + b\mathbf{u} = \mathbf{w}$.

2. Let $\mathbf{u} = \langle 2, 1 \rangle$ and $\mathbf{v} = \langle 1, 2 \rangle$.

    (a) Determine the components and draw geometric representations of the vectors $2\mathbf{u}$, $\frac{1}{2}\mathbf{u}$, $(-1)\mathbf{u}$, and $(-3)\mathbf{u}$ on the same set of axes.

    (b) Determine the components and draw geometric representations of the vectors $\mathbf{u} + \mathbf{v}$, $\mathbf{u} + 2\mathbf{v}$, and $\mathbf{u} + 3\mathbf{v}$.

    (c) Determine the components and draw geometric representations of the vectors $\mathbf{u} - \mathbf{v}$, $\mathbf{u} - 2\mathbf{v}$, and $\mathbf{u} - 3\mathbf{v}$.

    (d) Recall that $\mathbf{u} - \mathbf{v} = \mathbf{u} + (-1)\mathbf{v}$. Use the "tip to tail" perspective for vector addition to explain why the difference $\mathbf{u} - \mathbf{v}$ can be viewed as a vector that points from the tip of $\mathbf{v}$ to the tip of $\mathbf{u}$.

3. Recall that given any vector $\mathbf{v}$, we can calculate its length, $|\mathbf{v}|$. Also, we say that two vectors that are scalar multiples of one another are *parallel*.

(a) Let $\mathbf{v} = \langle 3, 4 \rangle$ in $\mathbb{R}^2$. Compute $|\mathbf{v}|$, and determine the components of the vector $\mathbf{u} = \frac{1}{|\mathbf{v}|}\mathbf{v}$. What is the magnitude of the vector $\mathbf{u}$? How does its direction compare to $\mathbf{v}$?

(b) Let $\mathbf{w} = 3\mathbf{i} - 3\mathbf{j}$ in $\mathbb{R}^2$. Determine a unit vector $\mathbf{u}$ in the same direction as $\mathbf{w}$.

(c) Let $\mathbf{v} = \langle 2, 3, 5 \rangle$ in $\mathbb{R}^3$. Compute $|\mathbf{v}|$, and determine the components of the vector $\mathbf{u} = \frac{1}{|\mathbf{v}|}\mathbf{v}$. What is the magnitude of the vector $\mathbf{u}$? How does its direction compare to $\mathbf{v}$?

(d) Let $\mathbf{v}$ be an arbitrary nonzero vector in $\mathbb{R}^3$. Write a general formula for a unit vector that is parallel to $\mathbf{v}$.

9.3 The Dot Product

Motivating Questions

In this section, we strive to understand the ideas generated by the following important questions:

- How is the dot product of two vectors defined and what geometric information does it tell us?
- How can we tell if two vectors in $\mathbb{R}^n$ are perpendicular?
- How do we find the projection of one vector onto another?

Introduction

In the last section, we considered vector addition and scalar multiplication and found that each operation had a natural geometric interpretation. In this section, we will introduce a means of multiplying vectors.

HW due 02.26.18

Preview Activity 9.3. For two-dimensional vectors $\mathbf{u} = \langle u_1, u_2 \rangle$ and $\mathbf{v} = \langle v_1, v_2 \rangle$, the dot product is simply the scalar obtained by:
$$\mathbf{u} \cdot \mathbf{v} = u_1 v_1 + u_2 v_2.$$

(a) If $\mathbf{u} = \langle 3, 4 \rangle$ and $\mathbf{v} = \langle -2, 1 \rangle$, find the dot product $\mathbf{u} \cdot \mathbf{v}$.

(b) Find $\mathbf{i} \cdot \mathbf{i}$ and $\mathbf{i} \cdot \mathbf{j}$.

(c) If $\mathbf{u} = \langle 3, 4 \rangle$, find $\mathbf{u} \cdot \mathbf{u}$. How is this related to $|\mathbf{u}|$?

(d) On the axes in Figure 9.34, plot the vectors $\mathbf{u} = \langle 1, 3 \rangle$ and $\mathbf{v} = \langle -3, 1 \rangle$. Then, find $\mathbf{u} \cdot \mathbf{v}$. What is the angle between these vectors?

use trig

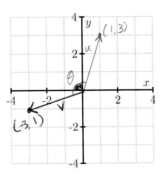

Figure 9.34: For part (d)

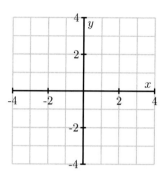

Figure 9.35: For part (e)

(e) On the axes in Figure 9.35, plot the vector $\mathbf{u} = \langle 1, 3 \rangle$.

For each of the following vectors $\mathbf{v}$, plot the vector on Figure 9.35 and then compute the dot product $\mathbf{u} \cdot \mathbf{v}$.

- $\mathbf{v} = \langle 3, 2 \rangle$.
- $\mathbf{v} = \langle 3, 0 \rangle$.
- $\mathbf{v} = \langle 3, -1 \rangle$.
- $\mathbf{v} = \langle 3, -2 \rangle$.
- $\mathbf{v} = \langle 3, -4 \rangle$.

(f) Based upon the previous part of this activity, what do you think is the sign of the dot product in the following three cases shown in Figure 9.36?

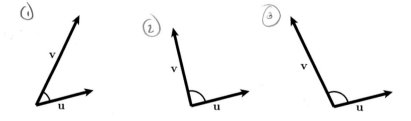

Figure 9.36: For part (f)

9.3. THE DOT PRODUCT

The Dot Product

If we have two n-dimensional vectors $\mathbf{u} = \langle u_1, u_2, \ldots, u_n \rangle$ and $\mathbf{v} = \langle v_1, v_2, \ldots, v_n \rangle$, we define their *dot product*[6] as

$$\mathbf{u} \cdot \mathbf{v} = u_1 v_1 + u_2 v_2 + \ldots + u_n v_n.$$

For instance, we find that

$$\langle 3, 0, 1 \rangle \cdot \langle -2, 1, 4 \rangle = 3 \cdot (-2) + 0 \cdot 1 + 1 \cdot 4 = -6 + 0 + 4 = -2.$$

Notice that the resulting quantity is a scalar. Our work in Preview Activity 9.3 examined dot products of two-dimensional vectors.

Activity 9.12.

Determine each of the following.

(a) $\langle 1, 2, -3 \rangle \cdot \langle 4, -2, 0 \rangle$.

(b) $\langle 0, 3, -2, 1 \rangle \cdot \langle 5, -6, 0, 4 \rangle$

◁

The dot product is a natural way to define a product of two vectors. In addition, it behaves in ways that are similar to the product of, say, real numbers.

Let $\mathbf{u}$, $\mathbf{v}$, and $\mathbf{w}$ be vectors in $\mathbb{R}^n$. Then

(a) $\mathbf{u} \cdot \mathbf{v} = \mathbf{v} \cdot \mathbf{u}$ (the dot product is *commutative*), and

(b) if c is a scalar, then $(c\mathbf{u} + \mathbf{v}) \cdot \mathbf{w} = c(\mathbf{u} \cdot \mathbf{w}) + \mathbf{v} \cdot \mathbf{w}$ (the dot product is *bilinear*),

---

Moreover, the dot product gives us valuable geometric information about the vectors and their relative orientation. For instance, let's consider what happens when we dot a vector with itself:

$$\mathbf{u} \cdot \mathbf{u} = \langle u_1, u_2, \ldots, u_n \rangle \cdot \langle u_1, u_2, \ldots, u_n \rangle = u_1^2 + u_2^2 + \ldots + u_n^2 = |\mathbf{u}|^2.$$

In other words, the dot product of a vector with itself gives the square of the length of the vector: $\mathbf{u} \cdot \mathbf{u} = |\mathbf{u}|^2$.

The angle between vectors

The dot product can help us understand the angle between two vectors. For instance, if we are given two vectors $\mathbf{u}$ and $\mathbf{v}$, there are two angles that these vectors create, as depicted in Figure 9.37. We will call θ, the smaller of these angles, the *angle between these vectors*. Notice that θ lies between 0 and π.

---

[6] As we will see shortly, the dot product arises in physics to calculate the work done by a vector force in a given direction. It might be more natural to define the dot product in this context, but it is more convenient from a mathematical perspective to define the dot product algebraically and then view work as an application of this definition.

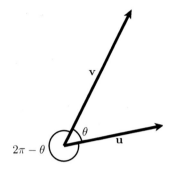

Figure 9.37: The angle between vectors u and v.

Figure 9.38: The triangle formed by u, v, and u − v.

To determine this angle, we may apply the Law of Cosines to the triangle shown in Figure 9.38.

Using the fact that the dot product of a vector with itself gives us the square of its length, together with the bilinearity of the dot product, we find:

$$\begin{aligned} |\mathbf{u} - \mathbf{v}|^2 &= |\mathbf{u}|^2 + |\mathbf{v}|^2 - 2|\mathbf{u}||\mathbf{v}|\cos(\theta) \\ (\mathbf{u} - \mathbf{v}) \cdot (\mathbf{u} - \mathbf{v}) &= \mathbf{u} \cdot \mathbf{u} + \mathbf{v} \cdot \mathbf{v} - 2|\mathbf{u}||\mathbf{v}|\cos(\theta) \\ \mathbf{u} \cdot (\mathbf{u} - \mathbf{v}) - \mathbf{v} \cdot (\mathbf{u} - \mathbf{v}) &= \mathbf{u} \cdot \mathbf{u} + \mathbf{v} \cdot \mathbf{v} - 2|\mathbf{u}||\mathbf{v}|\cos(\theta) \\ \mathbf{u} \cdot \mathbf{u} - 2\mathbf{u} \cdot \mathbf{v} + \mathbf{v} \cdot \mathbf{v} &= \mathbf{u} \cdot \mathbf{u} + \mathbf{v} \cdot \mathbf{v} - 2|\mathbf{u}||\mathbf{v}|\cos(\theta) \\ -2\mathbf{u} \cdot \mathbf{v} &= -2|\mathbf{u}||\mathbf{v}|\cos(\theta) \\ \mathbf{u} \cdot \mathbf{v} &= |\mathbf{u}||\mathbf{v}|\cos(\theta). \end{aligned}$$

To summarize, we have the important relationship

$$\mathbf{u} \cdot \mathbf{v} = u_1 v_1 + u_2 v_2 + \ldots + u_n v_n = |\mathbf{u}||\mathbf{v}|\cos(\theta). \tag{9.3}$$

It is sometimes useful to think of Equation (9.3) as giving us an expression for the angle between two vectors:

$$\theta = \cos^{-1}\left(\frac{\mathbf{u} \cdot \mathbf{v}}{|\mathbf{u}||\mathbf{v}|}\right).$$

The real beauty of this expression is this: the dot product is a very simple algebraic operation to perform yet it provides us with important geometric information – namely the angle between the vectors – that would be difficult to determine otherwise.

Activity 9.13.

Determine each of the following.

(a) The length of the vector $\mathbf{u} = \langle 1, 2, -3 \rangle$ using the dot product.

(b) The angle between the vectors $\mathbf{u} = \langle 1, 2 \rangle$ and $\mathbf{v} = \langle 4, -1 \rangle$ to the nearest tenth of a degree.

9.3. THE DOT PRODUCT

(c) The angle between the vectors $\mathbf{y} = \langle 1, 2, -3 \rangle$ and $\mathbf{z} = \langle -2, 1, 1 \rangle$ to the nearest tenth of a degree.

(d) If the angle between the vectors $\mathbf{u}$ and $\mathbf{v}$ is a right angle, what does the expression $\mathbf{u} \cdot \mathbf{v} = |\mathbf{u}||\mathbf{v}| \cos \theta$ say about their dot product?

(e) If the angle between the vectors $\mathbf{u}$ and $\mathbf{v}$ is acute—that is, less than $\pi/2$—what does the expression $\mathbf{u} \cdot \mathbf{v} = |\mathbf{u}||\mathbf{v}| \cos \theta$ say about their dot product?

(f) If the angle between the vectors $\mathbf{u}$ and $\mathbf{v}$ is obtuse—that is, greater than $\pi/2$—what does the expression $\mathbf{u} \cdot \mathbf{v} = |\mathbf{u}||\mathbf{v}| \cos \theta$ say about their dot product?

◁

The Dot Product and Orthogonality

When the angle between two vectors is a right angle, it is frequently the case that something important is happening. In this case, we say the vectors are *orthogonal*. For instance, orthogonality often plays a role in optimization problems; to determine the shortest path from a point in $\mathbb{R}^3$ to a given plane, we move along a line orthogonal to the plane.

As Activity 9.13 indicates, the dot product provides a simple means to determine whether two vectors are orthogonal to one another. In this case, $\mathbf{u} \cdot \mathbf{v} = |\mathbf{u}||\mathbf{v}| \cos(\pi/2) = 0$, so we make the following important observation.

> Two vectors $\mathbf{u}$ and $\mathbf{v}$ in $\mathbb{R}^n$ are orthogonal to each other if $\mathbf{u} \cdot \mathbf{v} = 0$.

More generally, the sign of the dot product gives us useful information about the relative orientation of the vectors. If we remember that

$$\cos(\theta) > 0 \quad \text{if } \theta \text{ is an acute angle,}$$
$$\cos(\theta) = 0 \quad \text{if } \theta \text{ is a right angle,}$$
$$\text{and } \cos(\theta) < 0 \quad \text{if } \theta \text{ is an obtuse angle,}$$

we see that

$$\mathbf{u} \cdot \mathbf{v} > 0 \quad \text{if } \theta \text{ is an acute angle,}$$
$$\mathbf{u} \cdot \mathbf{v} = 0 \quad \text{if } \theta \text{ is a right angle,}$$
$$\text{and } \mathbf{u} \cdot \mathbf{v} < 0 \quad \text{if } \theta \text{ is an obtuse angle.}$$

This is illustrated in Figure 9.39.

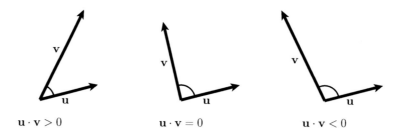

Figure 9.39: The orientation of vectors

Work, Force, and Displacement

In physics, work is a measure of the energy required to apply a force to an object through a displacement. For instance, Figure 9.40 shows a force **F** displacing an object from point A to point B. The displacement is then represented by the vector $\overrightarrow{AB}$.

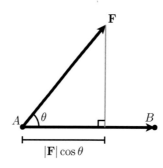

Figure 9.40: A force **F** displacing an object.

It turns out that the work required to displace the object is

$$W = \mathbf{F} \cdot \overrightarrow{AB} = |\mathbf{F}||\overrightarrow{AB}|\cos(\theta).$$

This means that the work is determined only by the magnitude of the force applied parallel to the displacement. Consequently, if we are given two vectors **u** and **v**, we would like to write **u** as a sum of two vectors, one of which is parallel to **v** and one of which is orthogonal to **v**. We take up this task after the next activity.

Activity 9.14.

Determine the work done by a 25 pound force acting at a 30° angle to the direction of the object's motion, if the object is pulled 10 feet. In addition, is more work or less work done if the angle to the direction of the object's motion is 60°?

◁

Projections

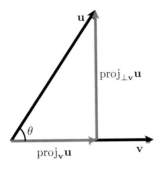

Figure 9.41: The projection of **u** onto **v**.

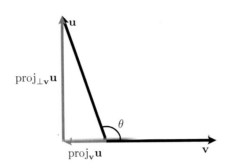

Figure 9.42: $\text{proj}_\mathbf{v} \mathbf{u}$ when $\theta > \frac{\pi}{2}$.

Suppose we are given two vectors **u** and **v** as shown in Figure 9.41. Motivated by our discussion of work, we would like to write **u** as a sum of two vectors, one of which is parallel to **v** and one of which is orthogonal. That is, we would like to write

$$\mathbf{u} = \text{proj}_\mathbf{v} \mathbf{u} + \text{proj}_{\perp \mathbf{v}} \mathbf{u}, \tag{9.4}$$

where $\text{proj}_\mathbf{v} \mathbf{u}$ is parallel to **v** and $\text{proj}_{\perp \mathbf{v}} \mathbf{u}$ is orthogonal to **v**. We call the vector $\text{proj}_\mathbf{v} \mathbf{u}$ the *projection of* **u** *onto* **v**.

To find the vector $\text{proj}_\mathbf{v} \mathbf{u}$, we will dot both sides of Equation (9.4) with the vector **v**, to find that

$$\begin{aligned} \mathbf{u} \cdot \mathbf{v} &= (\text{proj}_\mathbf{v} \mathbf{u} + \text{proj}_{\perp \mathbf{v}} \mathbf{u}) \cdot \mathbf{v} \\ &= (\text{proj}_\mathbf{v} \mathbf{u}) \cdot \mathbf{v} + (\text{proj}_{\perp \mathbf{v}} \mathbf{u}) \cdot \mathbf{v} \\ &= (\text{proj}_\mathbf{v} \mathbf{u}) \cdot \mathbf{v}. \end{aligned}$$

Notice that $(\text{proj}_{\perp \mathbf{v}} \mathbf{u}) \cdot \mathbf{v} = 0$ since $\text{proj}_{\perp \mathbf{v}} \mathbf{u}$ is orthogonal to **v**. Also, $\text{proj}_\mathbf{v} \mathbf{u}$ must be a scalar multiple of **v** since it is parallel to **v**, so we will write $\text{proj}_\mathbf{v} \mathbf{u} = s\mathbf{v}$. It follows that

$$\mathbf{u} \cdot \mathbf{v} = (\text{proj}_\mathbf{v} \mathbf{u}) \cdot \mathbf{v} = s\mathbf{v} \cdot \mathbf{v},$$

which means that

$$s = \frac{\mathbf{u} \cdot \mathbf{v}}{\mathbf{v} \cdot \mathbf{v}}$$

and hence

$$\text{proj}_\mathbf{v} \mathbf{u} = \frac{\mathbf{u} \cdot \mathbf{v}}{\mathbf{v} \cdot \mathbf{v}} \mathbf{v} = \frac{\mathbf{u} \cdot \mathbf{v}}{|\mathbf{v}|^2} \mathbf{v}$$

It is sometimes useful to write $\text{proj}_\mathbf{v} \mathbf{u}$ as a scalar times a unit vector in the direction of **v**. We call this scalar the *component of* **u** *along* **v** and denote it as $\text{comp}_\mathbf{v} \mathbf{u}$. We therefore have

$$\text{proj}_\mathbf{v} \mathbf{u} = \frac{\mathbf{u} \cdot \mathbf{v}}{|\mathbf{v}|^2} \mathbf{v} = \frac{\mathbf{u} \cdot \mathbf{v}}{|\mathbf{v}|} \frac{\mathbf{v}}{|\mathbf{v}|} = \text{comp}_\mathbf{v} \mathbf{u} \frac{\mathbf{v}}{|\mathbf{v}|},$$

so that
$$\text{comp}_{\mathbf{v}} \mathbf{u} = \frac{\mathbf{u} \cdot \mathbf{v}}{|\mathbf{v}|}.$$

Let $\mathbf{u}$ and $\mathbf{v}$ be vectors in $\mathbb{R}^n$. The component of $\mathbf{u}$ in the direction of $\mathbf{v}$ is the scalar
$$\text{comp}_{\mathbf{v}} \mathbf{u} = \frac{\mathbf{u} \cdot \mathbf{v}}{|\mathbf{v}|},$$
and the projection of $\mathbf{u}$ onto $\mathbf{v}$ is the vector
$$\text{proj}_{\mathbf{v}} \mathbf{u} = \frac{\mathbf{u} \cdot \mathbf{v}}{\mathbf{v} \cdot \mathbf{v}} \mathbf{v}.$$

Moreover, since
$$\mathbf{u} = \text{proj}_{\mathbf{v}} \mathbf{u} + \text{proj}_{\perp \mathbf{v}} \mathbf{u},$$
it follows that
$$\text{proj}_{\perp \mathbf{v}} \mathbf{u} = \mathbf{u} - \text{proj}_{\mathbf{v}} \mathbf{u}.$$
This shows that once we have computed $\text{proj}_{\mathbf{v}} \mathbf{u}$, we can find $\text{proj}_{\perp \mathbf{v}} \mathbf{u}$ simply by calculating the difference of two known vectors.

Activity 9.15.

Let $\mathbf{u} = \langle 2, 6 \rangle$ and $\mathbf{v} = \langle 4, -8 \rangle$. Find $\text{comp}_{\mathbf{v}} \mathbf{u}$, $\text{proj}_{\mathbf{v}} \mathbf{u}$ and $\text{proj}_{\perp \mathbf{v}} \mathbf{u}$, and draw a picture to illustrate. Finally, express $\mathbf{u}$ as the sum of two vectors where one is parallel to $\mathbf{v}$ and the other is perpendicular to $\mathbf{v}$.

◁

Summary

- The dot product of two vectors in $\mathbb{R}^n$, $\mathbf{u} = \langle u_1, u_2, \ldots, u_n \rangle$ and $\mathbf{v} = \langle v_1, v_2, \ldots, v_n \rangle$, is the scalar
$$\mathbf{u} \cdot \mathbf{v} = u_1 v_1 + u_2 v_2 + \cdots + u_n v_n.$$

- The dot product is related to the length of a vector since $\mathbf{u} \cdot \mathbf{u} = |\mathbf{u}|^2$.

- The dot product provides us with information about the angle between the vectors since
$$\mathbf{u} \cdot \mathbf{v} = |\mathbf{u}| \, |\mathbf{v}| \cos(\theta),$$
where θ is the angle between $\mathbf{u}$ and $\mathbf{v}$.

- Two vectors are orthogonal if the angle between them is $\pi/2$. In terms of the dot product, the vectors $\mathbf{u}$ and $\mathbf{v}$ are orthogonal if and only if $\mathbf{u} \cdot \mathbf{v} = 0$.

- The projection of a vector $\mathbf{u}$ in $\mathbb{R}^n$ onto a vector $\mathbf{v}$ in $\mathbb{R}^n$ is the vector
$$\text{proj}_{\mathbf{v}} \mathbf{u} = \frac{\mathbf{u} \cdot \mathbf{v}}{\mathbf{v} \cdot \mathbf{v}} \mathbf{v}.$$

9.3. THE DOT PRODUCT

Exercises

1. Let $\mathbf{v} = \langle -2, 5 \rangle$ in $\mathbb{R}^2$, and let $\mathbf{y} = \langle 0, 3, -2 \rangle$ in $\mathbb{R}^3$.

    (a) Is $\langle 2, -1 \rangle$ perpendicular to $\mathbf{v}$? Why or why not?

    (b) Find a unit vector $\mathbf{u}$ in $\mathbb{R}^2$ such that $\mathbf{u}$ is perpendicular to $\mathbf{v}$. How many such vectors are there?

    (c) Is $\langle 2, -1, -2 \rangle$ perpendicular to $\mathbf{y}$? Why or why not?

    (d) Find a unit vector $\mathbf{w}$ in $\mathbb{R}^3$ such that $\mathbf{w}$ is perpendicular to $\mathbf{y}$. How many such vectors are there?

    (e) Let $\mathbf{z} = \langle 2, 1, 0 \rangle$. Find a unit vector $\mathbf{r}$ in $\mathbb{R}^3$ such that $\mathbf{r}$ is perpendicular to both $\mathbf{y}$ and $\mathbf{z}$. How many such vectors are there?

2. Consider the triangle in $\mathbb{R}^3$ given by $P(3, 2, -1)$, $Q(1, -2, 4)$, and $R(4, 4, 0)$.

    (a) Find the measure of each of the three angles in the triangle, accurate to 0.01 degrees.

    (b) Choose two sides of the triangle, and call the vectors that form the sides (emanating from a common point) $\mathbf{a}$ and $\mathbf{b}$.

    i. Compute $\text{proj}_\mathbf{b} \mathbf{a}$, and $\text{proj}_{\perp \mathbf{b}} \mathbf{a}$.

    ii. Explain why $\text{proj}_{\perp \mathbf{b}} \mathbf{a}$ can be considered a height of triangle PQR.

    iii. Find the area of the given triangle.

3. Let $\mathbf{u}$ and $\mathbf{v}$ be vectors in $\mathbb{R}^5$ with $\mathbf{u} \cdot \mathbf{v} = -1$, $|\mathbf{u}| = 2$, $|\mathbf{v}| = 3$, and θ the angle between $\mathbf{u}$ and $\mathbf{v}$. Use the properties of the dot product to find each of the following.

    (a) $\mathbf{u} \cdot 2\mathbf{v}$

    (b) $(\mathbf{u} + \mathbf{v}) \cdot \mathbf{v}$

    (c) $(2\mathbf{u} + 4\mathbf{v}) \cdot (\mathbf{u} - 7\mathbf{v})$

    (d) $\mathbf{v} \cdot \mathbf{v}$

    (e) $|\mathbf{u}||\mathbf{v}| \cos(\theta)$

    (f) θ

9.4 The Cross Product

Motivating Questions

In this section, we strive to understand the ideas generated by the following important questions:

- How and when is the cross product of two vectors defined?
- What geometric information does the cross product provide?

The last two sections have introduced some basic algebraic operations on vectors—addition, scalar multiplication, and the dot product—with useful geometric interpretations. In this section, we will meet a final algebraic operation, the *cross product*, which again conveys important geometric information.

To begin, we must emphasize that the cross product is only defined for vectors **u** and **v** in $\mathbb{R}^3$. Also, remember that we use a right-hand coordinate system, as described in Section 9.1. In particular, recall that the vectors **i**, **j**, and **k** are oriented as shown below in Figure 9.43. Earlier, we noticed that if we point the index finger of our *right* hand in the direction of **i** and our middle finger in the direction of **j**, then our thumb points in the direction of **k**.

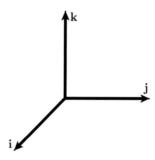

Figure 9.43: Basis vectors **i**, **j**, and **k**.

Preview Activity 9.4. The cross product of two vectors, **u** and **v**, will itself be a *vector* denoted **u** × **v**. The direction of **u** × **v** is determined by the right-hand rule: if we point the index finger of our right hand in the direction of **u** and our middle finger in the direction of **v**, then our thumb points in the direction of **u** × **v**.

(a) We begin by defining the cross products using the vectors **i**, **j**, and **k**. Referring to Figure 9.43, explain why **i**, **j**, **k** in that order form a right-hand system. We then define **i** × **j** to be **k** – that is **i** × **j** = **k**.

(b) Now explain why **i**, **k**, and −**j** in that order form a right-hand system. We then define **i** × **k** to be −**j** – that is **i** × **k** = −**j**.

(c) Continuing in this way, complete the missing entries in Table 9.4.

9.4. THE CROSS PRODUCT

$i \times j = k$	$i \times k = -j$	$j \times k =$
$j \times i =$	$k \times i =$	$k \times j =$

Table 9.4: Table of cross products involving i, j, and k.

the same but neg.

(d) Up to this point, the products you have seen, such as the product of real numbers and the dot product of vectors, have been commutative, meaning that the product does not depend on the order of the terms. For instance, $2 \cdot 5 = 5 \cdot 2$. The table above suggests, however, that the cross product is *anti-commutative*: for any vectors u and v in $\mathbb{R}^3$, $u \times v = -v \times u$. If we consider the case when $u = v$, this shows that $v \times v = -v \times v$. What does this tell us about $v \times v$; in particular, what vector is unchanged by scalar multiplication by -1?

(e) The cross product is also a *bilinear* operation, meaning that it interacts with scalar multiplication and vector addition as one would expect: $(cu + v) \times w = c(u \times w) + v \times w$. For example,

$$(2i + j) \times k = 2(i \times k) + (j \times k) = -2j + i.$$

$\langle 0, -2, 0 \rangle + \langle 1, 0, 0 \rangle = \langle 1, -2, 0 \rangle$

Using this property along with Table 9.4, find the cross product $u \times v$ if $u = 2i + 3j$ and $v = -i + k$.

(f) Verify that the cross product $u \times v$ you just found in part (e) is orthogonal to both u and v.

(g) Consider the vectors u and v in the xy-plane as shown below in Figure 9.44.

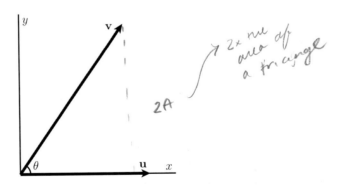

Figure 9.44: Two vectors in the xy-plane

Explain why $u = |u|i$ and $v = |v|\cos(\theta)i + |v|\sin(\theta)j$. Then compute the length of $|u \times v|$.

(h) Multiplication of real numbers is *associative*, which means, for instance, that $(2 \cdot 5) \cdot 3 = 2 \cdot (5 \cdot 3)$. Is it true that the cross product of vectors is associative? For instance, is it true that $(i \times j) \times j = i \times (j \times j)$?

Computing the cross product

As we have seen in Preview Activity 9.4, the cross product $\mathbf{u} \times \mathbf{v}$ is defined for two vectors $\mathbf{u}$ and $\mathbf{v}$ in $\mathbb{R}^3$ and produces another vector in $\mathbb{R}^3$. Using the right-hand rule, we saw that

$$\mathbf{i} \times \mathbf{j} = \mathbf{k}, \qquad \mathbf{i} \times \mathbf{k} = -\mathbf{j}, \qquad \mathbf{j} \times \mathbf{k} = \mathbf{i}$$

$$\mathbf{j} \times \mathbf{i} = -\mathbf{k} \qquad \mathbf{k} \times \mathbf{i} = \mathbf{j} \qquad \mathbf{k} \times \mathbf{j} = -\mathbf{i}$$

If in addition we apply the bilinearity property to compute the cross product in terms of the components of general vectors, $\mathbf{u}$ and $\mathbf{v}$ we see that[7]

$$\begin{aligned}
\mathbf{u} \times \mathbf{v} &= (u_1 \mathbf{i} + u_2 \mathbf{j} + u_3 \mathbf{k}) \times (v_1 \mathbf{i} + v_2 \mathbf{j} + v_3 \mathbf{k}) \\
&= u_1 \mathbf{i} \times (v_1 \mathbf{i} + v_2 \mathbf{j} + v_3 \mathbf{k}) + \\
&\quad u_2 \mathbf{j} \times (v_1 \mathbf{i} + v_2 \mathbf{j} + v_3 \mathbf{k}) + \\
&\quad u_3 \mathbf{k} \times (v_1 \mathbf{i} + v_2 \mathbf{j} + v_3 \mathbf{k}) \\
&= u_1 v_1 \mathbf{i} \times \mathbf{i} + u_1 v_2 \mathbf{i} \times \mathbf{j} + u_1 v_3 \mathbf{i} \times \mathbf{k} + \\
&\quad u_2 v_1 \mathbf{j} \times \mathbf{i} + u_2 v_2 \mathbf{j} \times \mathbf{j} + u_2 v_3 \mathbf{j} \times \mathbf{k} + \\
&\quad u_3 v_1 \mathbf{k} \times \mathbf{i} + u_3 v_2 \mathbf{k} \times \mathbf{j} + u_3 v_3 \mathbf{k} \times \mathbf{k} \\
&= u_1 v_2 \mathbf{k} - u_1 v_3 \mathbf{j} - u_2 v_1 \mathbf{k} + u_2 v_3 \mathbf{i} + u_3 v_1 \mathbf{j} - u_3 v_2 \mathbf{i} \\
&= (u_2 v_3 - u_3 v_2) \mathbf{i} - (u_1 v_3 - u_3 v_1) \mathbf{j} + (u_1 v_2 - u_2 v_1) \mathbf{k}
\end{aligned}$$

To summarize, we have shown that

$$(u_1 \mathbf{i} + u_2 \mathbf{j} + u_3 \mathbf{k}) \times (v_1 \mathbf{i} + v_2 \mathbf{j} + v_3 \mathbf{k}) = (u_2 v_3 - u_3 v_2) \mathbf{i} - (u_1 v_3 - u_3 v_1) \mathbf{j} + (u_1 v_2 - u_2 v_1) \mathbf{k}. \tag{9.5}$$

At first, this may look intimidating and difficult to remember. However, if we rewrite the expression in Equation (9.5) using determinants, important structure emerges. The determinant of a 2×2 matrix is

$$\begin{vmatrix} a & b \\ c & d \end{vmatrix} = ad - bc.$$

It follows that we can thus rewrite Equation (9.5) in the form

$$\mathbf{u} \times \mathbf{v} = \begin{vmatrix} u_2 & u_3 \\ v_2 & v_3 \end{vmatrix} \mathbf{i} - \begin{vmatrix} u_1 & u_3 \\ v_1 & v_3 \end{vmatrix} \mathbf{j} + \begin{vmatrix} u_1 & u_2 \\ v_1 & v_2 \end{vmatrix} \mathbf{k}.$$

For those familiar with the determinant of a 3×3 matrix, we write the mnemonic

$$\mathbf{u} \times \mathbf{v} = \begin{vmatrix} \mathbf{i} & \mathbf{j} & \mathbf{k} \\ u_1 & u_2 & u_3 \\ v_1 & v_2 & v_3 \end{vmatrix}.$$

---

[7] Like the dot product, the cross product arises in physical applications, e.g., torque, but is it more convenient mathematically to begin from an algebraic perspective.

9.4. THE CROSS PRODUCT

Activity 9.16.

Suppose $\mathbf{u} = \langle 2, -1, 0 \rangle$ and $\mathbf{v} = \langle 0, 1, 3 \rangle$. Use the formula (9.5) for the following.

(a) Find the cross product $\mathbf{u} \times \mathbf{v}$.

(b) Find the cross product $\mathbf{u} \times \mathbf{i}$.

(c) Find the cross product $\mathbf{u} \times \mathbf{u}$.

(d) Evaluate the dot products $\mathbf{u} \cdot (\mathbf{u} \times \mathbf{v})$ and $\mathbf{v} \cdot (\mathbf{u} \times \mathbf{v})$. What does this tell you about the geometric relationship among $\mathbf{u}$, $\mathbf{v}$, and $\mathbf{u} \times \mathbf{v}$?

◁

The cross product satisfies the following properties, which were illustrated in Preview Activity 9.4 and may be easily verified from the definition (9.5).

Let $\mathbf{u}$, $\mathbf{v}$, and $\mathbf{w}$ be vectors in $\mathbb{R}^3$, and let c be a scalar. Then

1. $\mathbf{u} \times \mathbf{v} = -(\mathbf{v} \times \mathbf{u})$

2. $(c\mathbf{u} + \mathbf{v}) \times \mathbf{w} = c(\mathbf{u} \times \mathbf{w}) + \mathbf{v} \times \mathbf{w}$

3. $\mathbf{u} \times \mathbf{v} = \mathbf{0}$ if $\mathbf{u}$ and $\mathbf{v}$ are parallel.

4. The cross product is not associative; that is, in general
$$(\mathbf{u} \times \mathbf{v}) \times \mathbf{w} \neq \mathbf{u} \times (\mathbf{v} \times \mathbf{w}).$$

Just as we found for the dot product, the cross product provides us with useful geometric information. In particular, both the length and direction of the cross product $\mathbf{u} \times \mathbf{v}$ encode information about the geometric relationship between $\mathbf{u}$ and $\mathbf{v}$.

The Length of $\mathbf{u} \times \mathbf{v}$

We may ask whether the length $|\mathbf{u} \times \mathbf{v}|$ has any relationship to the lengths of $\mathbf{u}$ and $\mathbf{v}$. To investigate, we will compute the square of the length $|\mathbf{u} \times \mathbf{v}|^2$ and denote by θ the angle between $\mathbf{u}$ and $\mathbf{v}$, as in Section 9.3. Doing so, we find through some significant algebra that

$$\begin{aligned}
|\mathbf{u} \times \mathbf{v}|^2 &= (u_2v_3 - u_3v_2)^2 + (u_1v_3 - u_3v_1)^2 + (u_1v_2 - u_2v_1)^2 \\
&= u_2^2v_3^2 - 2u_2u_3v_2v_3 + u_3^2v_2^2 + u_1^2v_3^2 - 2u_1u_3v_1v_3 + u_3^2v_1^2 + u_1^2v_2^2 - 2u_1u_2v_1v_2 + u_2^2v_1^2 \\
&= u_1^2(v_2^2 + v_3^2) + u_2^2(v_1^2 + v_3^2) + u_3^2(v_1^2 + v_2^2) - 2(u_1u_2v_1v_2 + u_1u_3v_1v_3 + u_2u_3v_2v_3) \\
&= u_1^2(v_1^2 + v_2^2 + v_3^2) + u_2^2(v_1^2 + v_2^2 + v_3^2) + u_3^2(v_1^2 + v_2^2 + v_3^2) - \\
&\quad (u_1^2v_1^2 + u_2^2v_2^2 + u_3^2v_3^2 + 2(u_1u_2v_1v_2 + u_1u_3v_1v_3 + u_2u_3v_2v_3)) \\
&= (u_1^2 + u_2^2 + u_3^2)(v_1^2 + v_2^2 + v_3^2) - (u_1v_1 + u_2v_2 + u_3v_3)^2 \\
&= |\mathbf{u}|^2|\mathbf{v}|^2 - (\mathbf{u} \cdot \mathbf{v})^2 \\
&= |\mathbf{u}|^2|\mathbf{v}|^2(1 - \cos^2(\theta)) \\
&= |\mathbf{u}|^2|\mathbf{v}|^2 \sin^2(\theta).
\end{aligned}$$

Therefore, we have found $|\mathbf{u} \times \mathbf{v}|^2 = |\mathbf{u}|^2|\mathbf{v}|^2 \sin^2(\theta)$, which means that

$$|\mathbf{u} \times \mathbf{v}| = |\mathbf{u}||\mathbf{v}|\sin(\theta). \tag{9.6}$$

Note that the third property stated above says that $\mathbf{u} \times \mathbf{v} = 0$ if $\mathbf{u}$ and $\mathbf{v}$ are parallel. This is reflected in Equation (9.6) since $\sin(\theta) = 0$ if $\mathbf{u}$ and $\mathbf{v}$ are parallel, which implies that $\mathbf{u} \times \mathbf{v} = 0$.

Equation (9.6) also has a geometric implication. Consider the parallelogram formed by two vectors $\mathbf{u}$ and $\mathbf{v}$, as shown in Figure 9.45.

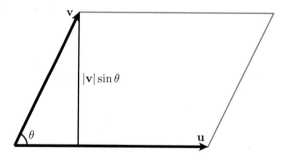

Figure 9.45: The parallelogram formed by $\mathbf{u}$ and $\mathbf{v}$

Remember that the area of a parallelogram is the product of its base and height. As shown in the figure, we may consider the base of the parallelogram to be $|\mathbf{u}|$ and the height to be $|\mathbf{v}|\sin(\theta)$. This means that the area of the parallelogram formed by $\mathbf{u}$ and $\mathbf{v}$ is

$$|\mathbf{u}||\mathbf{v}|\sin(\theta) = |\mathbf{u} \times \mathbf{v}|.$$

This leads to the following interesting fact.

> The length, $|\mathbf{u} \times \mathbf{v}|$, of the cross product of vectors $\mathbf{u}$ and $\mathbf{v}$ is the area of the parallelogram determined by $\mathbf{u}$ and $\mathbf{v}$.

9.4. THE CROSS PRODUCT

Activity 9.17.

(a) Find the area of the parallelogram formed by the vectors $\mathbf{u} = \langle 1, 3, -2 \rangle$ and $\mathbf{v} = \langle 3, 0, 1 \rangle$.

(b) Find the area of the parallelogram in $\mathbb{R}^3$ whose vertices are $(1, 0, 1)$, $(0, 0, 1)$, $(2, 1, 0)$, and $(1, 1, 0)$.

The Direction of $\mathbf{u} \times \mathbf{v}$

Now that we understand the length of $\mathbf{u} \times \mathbf{v}$, we will investigate its direction. Remember from Preview Activity 9.4 that cross products involving the vectors $\mathbf{i}$, $\mathbf{j}$, and $\mathbf{k}$ resulted in vectors that are orthogonal to the two terms. We will see that this holds more generally.

We begin by computing $\mathbf{u} \cdot (\mathbf{u} \times \mathbf{v})$, and see that

$$\begin{aligned} \mathbf{u} \cdot (\mathbf{u} \times \mathbf{v}) &= u_1(u_2 v_3 - u_3 v_2) - u_2(u_1 v_3 - u_3 v_1) + u_3(u_1 v_2 - u_2 v_1) \\ &= u_1 u_2 v_3 - u_1 u_3 v_2 - u_2 u_1 v_3 + u_2 u_3 v_1 + u_3 u_1 v_2 - u_3 u_2 v_1 \\ &= 0 \end{aligned}$$

To summarize, we have $\mathbf{u} \cdot (\mathbf{u} \times \mathbf{v}) = 0$, which implies that $\mathbf{u}$ is orthogonal to $\mathbf{u} \times \mathbf{v}$. In the same way, we can show that $\mathbf{v}$ is orthogonal to $\mathbf{u} \times \mathbf{v}$. The net effect is that $\mathbf{u} \times \mathbf{v}$ is a vector that is perpendicular to both $\mathbf{u}$ and $\mathbf{v}$, and hence $\mathbf{u} \times \mathbf{v}$ is perpendicular to the plane determined by $\mathbf{u}$ and $\mathbf{v}$. Moreover, the direction of $\mathbf{u} \times \mathbf{v}$ is determined by applying the right-hand rule to $\mathbf{u}$ and $\mathbf{v}$, as we saw in Preview Activity 9.4. In light of our earlier work that showed $|\mathbf{u}||\mathbf{v}|\sin(\theta) = |\mathbf{u} \times \mathbf{v}|$, we may now express $\mathbf{u} \times \mathbf{v}$ in the following different way.

> Suppose that $\mathbf{u}$ and $\mathbf{v}$ are not parallel and that $\mathbf{n}$ is the unit vector perpendicular to the plane containing $\mathbf{u}$ and $\mathbf{v}$ determined by the right-hand rule. Then
>
> $$\mathbf{u} \times \mathbf{v} = |\mathbf{u}||\mathbf{v}|\sin(\theta)\,\mathbf{n}.$$

There is yet one more geometric implication we may draw from this result. Suppose $\mathbf{u}$, $\mathbf{v}$, and $\mathbf{w}$ are vectors in $\mathbb{R}^3$ that are not coplanar and that form a three-dimension parallelepiped as shown in Figure 9.46.

The volume of the parallelepiped is determined by multiplying A, the area of the base, by the height h. As we have just seen, the area of the base is $|\mathbf{u} \times \mathbf{v}|$. Moreover, the height $h = |\mathbf{w}|\cos(\alpha)$ where α is the angle between $\mathbf{w}$ and the vector $\mathbf{n}$, which is orthogonal to the plane formed by $\mathbf{u}$ and $\mathbf{v}$. Since $\mathbf{n}$ is parallel to $\mathbf{u} \times \mathbf{v}$, the angle between $\mathbf{w}$ and $\mathbf{u} \times \mathbf{v}$ is also α. This shows that

$$|(\mathbf{u} \times \mathbf{v}) \cdot \mathbf{w}| = |\mathbf{u} \times \mathbf{w}||\mathbf{w}|\cos(\alpha) = Ah,$$

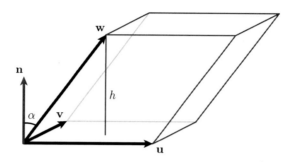

Figure 9.46: The parallelepiped determined by **u**, **v**, and **w**

and therefore

> The volume of the parallelepiped determined by **u**, **v**, and **w** is $|(\mathbf{u} \times \mathbf{v}) \cdot \mathbf{w}|$.

The quantity $|(\mathbf{u} \times \mathbf{v}) \cdot \mathbf{w}|$ is sometimes called the *scalar triple product*.

Activity 9.18.
Suppose $\mathbf{u} = \langle 3, 5, -1 \rangle$ and $\mathbf{v} = \langle 2, -2, 1 \rangle$.

(a) Find two unit vectors orthogonal to both **u** and **v**.

(b) Find the volume of the parallelepiped formed by the vectors **u**, **v**, and $\mathbf{w} = \langle 3, 3, 1 \rangle$.

(c) Find a vector orthogonal to the plane containing the points $(0, 1, 2)$, $(4, 1, 0)$, and $(-2, 2, 2)$.

(d) Given the vectors **u** and **v** shown below in Figure 9.47, sketch the cross products $\mathbf{u} \times \mathbf{v}$ and $\mathbf{v} \times \mathbf{u}$.

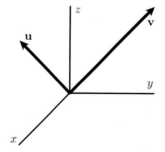

Figure 9.47: Vectors **u** and **v**

(e) Do the vectors $\mathbf{a} = \langle 1, 3, -2 \rangle$, $\mathbf{b} = \langle 2, 1, -4 \rangle$, and $\mathbf{c} = \langle 0, 1, 0 \rangle$ lie in the same plane? Use the concepts from this section to explain.

9.4. THE CROSS PRODUCT

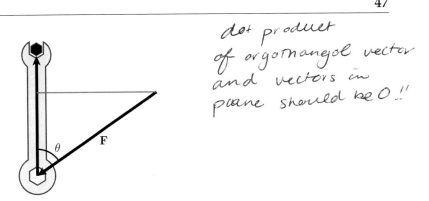

Figure 9.48. A force applied to a wrench

Torque is measured by a cross product

We have seen that the cross product enables us to produce a vector perpendicular to two given vectors, to measure the area of a parallelogram, and to measure the volume of a parallelepiped. Besides these geometric applications, the cross product also enables us to describe a physical quantity called *torque*.

Suppose that we would like to turn a bolt using a wrench as shown in Figure 9.48. If a force $\mathbf{F}$ is applied to the wrench and $\mathbf{r}$ is the vector from the position on the wrench at which the force is applied to center of the bolt, we define the torque, τ, to be

$$\tau = \mathbf{F} \times \mathbf{r}.$$

When a force is applied to an object, Newton's Second Law tells us that the force is equal to the rate of change of the object's linear momentum. Similarly, the torque applied to an object is equal to the rate of change of the object's angular momentum. In other words, torque will cause the bolt to rotate.

In many industrial applications, bolts are required to be tightened using a specified torque. Of course, the magnitude of the torque is $|\tau| = |\mathbf{F} \times \mathbf{r}| = |\mathbf{F}||\mathbf{r}|\sin(\theta)$. Thus, to produce a larger torque, we can increase either $|\mathbf{F}|$ or $|\mathbf{r}|$, which you may know if you have ever removed lug nuts when changing a flat tire. The ancient Greek mathematician Archimedes said: "Give me a lever long enough and a fulcrum on which to place it, and I shall move the world." A modern spin on this statement is: "Allow me to make $|\mathbf{r}|$ large enough, and I shall produce a torque large enough to move the world."

Comparing the dot and cross products

Finally, it is worthwhile to compare and contrast the dot and cross products.

(a) $\mathbf{u} \cdot \mathbf{v}$ is a scalar, while $\mathbf{u} \times \mathbf{v}$ is a vector.

(b) $\mathbf{u} \cdot \mathbf{v} = \mathbf{v} \cdot \mathbf{u}$, while $\mathbf{u} \times \mathbf{v} = -\mathbf{v} \times \mathbf{u}$

(c) $\mathbf{u} \cdot \mathbf{v} = |\mathbf{u}||\mathbf{v}|\cos(\theta)$, while $|\mathbf{u} \times \mathbf{v}| = |\mathbf{u}||\mathbf{v}|\sin(\theta)$.

(d) $\mathbf{u} \cdot \mathbf{v} = 0$ if $\mathbf{u}$ and $\mathbf{v}$ are perpendicular, while $\mathbf{u} \times \mathbf{v} = \mathbf{0}$ if $\mathbf{u}$ and $\mathbf{v}$ are parallel.

Summary

- The cross product is defined *only* for vectors in $\mathbb{R}^3$. The cross product of vectors $\mathbf{u}$ and $\mathbf{v}$ in $\mathbb{R}^3$ is the vector

$$(u_1\mathbf{i} + u_2\mathbf{j} + u_3\mathbf{k}) \times (v_1\mathbf{i} + v_2\mathbf{j} + v_3\mathbf{k}) = (u_2v_3 - u_3v_2)\mathbf{i} - (u_1v_3 - u_3v_1)\mathbf{j} + (u_1v_2 - u_2v_1)\mathbf{k}.$$

- Geometrically, the cross product is

$$\mathbf{u} \times \mathbf{v} = |\mathbf{u}|\,|\mathbf{v}|\,\sin(\theta)\,\mathbf{n},$$

where θ is the angle between $\mathbf{u}$ and $\mathbf{v}$ and $\mathbf{n}$ is a unit vector perpendicular to both $\mathbf{u}$ and $\mathbf{v}$ as determined by the right-hand rule.

- The cross product of vectors $\mathbf{u}$ and $\mathbf{v}$ is a vector perpendicular to both $\mathbf{u}$ and $\mathbf{v}$.

- The magnitude $|\mathbf{u} \times \mathbf{v}|$ of the cross product of the vectors $\mathbf{u}$ and $\mathbf{v}$ gives the area of the parallelogram determined by $\mathbf{u}$ and $\mathbf{v}$. Also, the scalar triple product $|(\mathbf{u} \times \mathbf{v}) \cdot \mathbf{w}|$ gives the volume of the parallelepiped determined by $\mathbf{u}$, $\mathbf{v}$, and $\mathbf{w}$.

Exercises

1. Let $\mathbf{u} = 2\mathbf{i} + \mathbf{j}$ and $\mathbf{v} = \mathbf{i} + 2\mathbf{j}$ be vectors in $\mathbb{R}^3$.

   (a) Without doing any computations, find a unit vector that is orthogonal to both $\mathbf{u}$ and $\mathbf{v}$. What does this tell you about the formula for $\mathbf{u} \times \mathbf{v}$?

   (b) Using the bilinearity of the cross product and what you know about cross products involving the fundamental vectors $\mathbf{i}$ and $\mathbf{j}$, compute $\mathbf{u} \times \mathbf{v}$.

   (c) Next, use the determinant version of Equation (9.5) to compute $\mathbf{u} \times \mathbf{v}$. Write one sentence that compares your results in (a), (b), and (c).

   (d) Find the area of the parallelogram determined by $\mathbf{u}$ and $\mathbf{v}$.

2. Let $\mathbf{x} = \langle 1, 1, 1 \rangle$ and $\mathbf{y} = \langle 0, 3, -2 \rangle$.

   (a) Are $\mathbf{x}$ and $\mathbf{y}$ orthogonal? Are $\mathbf{x}$ and $\mathbf{y}$ parallel? Clearly explain how you know, using appropriate vector products.

   (b) Find a unit vector that is orthogonal to both $\mathbf{x}$ and $\mathbf{y}$.

9.4. THE CROSS PRODUCT

(c) Express **y** as the sum of two vectors: one parallel to **x**, the other orthogonal to **x**.

(d) Determine the area of the parallelogram formed by **x** and **y**.

3. Consider the triangle in $\mathbb{R}^3$ formed by $P(3, 2, -1)$, $Q(1, -2, 4)$, and $R(4, 4, 0)$.

   (a) Find $\overrightarrow{PQ}$ and $\overrightarrow{PR}$.

   (b) Observe that the area of $\triangle PQR$ is half of the area of the parallelogram formed by $\overrightarrow{PQ}$ and $\overrightarrow{PR}$. Hence find the area of $\triangle PQR$.

   (c) Find a unit vector that is orthogonal to the plane that contains points P, Q, and R.

   (d) Determine the measure of $\angle PQR$.

9.5 Lines and Planes in Space

> **Motivating Questions**
>
> *In this section, we strive to understand the ideas generated by the following important questions:*
>
> - How are lines in $\mathbb{R}^3$ similar to and different from lines in $\mathbb{R}^2$?
> - What is the role that vectors play in representing equations of lines, particularly in $\mathbb{R}^3$?
> - How can we think of a plane as a set of points determined by a point and a vector?
> - How do we find the equation of a plane through three given non-collinear points?

Introduction

In single variable calculus, we learn that a differentiable function is *locally linear*. In other words, if we zoom in on the graph of a differentiable function at a point, the graph looks like the tangent line to the function at that point. In multivariable calculus, we will soon study curves in space; differentiable curves turn out to be locally linear as well. In addition, as we study functions of two variables, we will see that such a function is locally linear at a point if the surface defined by the function looks like a plane (the tangent plane) as we zoom in on the graph.

Consequently, it is important for us to understand both lines and planes in space. We begin our work by considering some familiar ideas in $\mathbb{R}^2$ but from a new perspective.

Preview Activity 9.5. We are familiar with equations of lines in the plane in the form $y = mx + b$, where m is the slope of the line and $(0, b)$ is the y-intercept. In this activity, we explore a more flexible way of representing lines that we can use not only in the plane, but in higher dimensions as well.

To begin, consider the line through the point $(2, -1)$ with slope $\frac{2}{3}$.

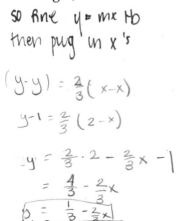

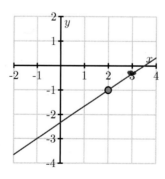

(a) Suppose we increase x by 1 from the point $(2, -1)$. How does the y-value change? What is the point on the line with x-coordinate 3?

9.5. LINES AND PLANES IN SPACE

(b) Suppose we decrease x by 3.25 from the point $(2, -1)$. How does the y-value change? What is the point on the line with x-coordinate -1.25?

(c) Now, suppose we increase x by some arbitrary value $3t$ [scalar] from the point $(2, -1)$. How does the y-value change? What is the point on the line with x-coordinate $2 + 3t$?

(d) Observe that the slope of the line is related to any vector whose y-component divided by the x-component is the slope of the line. For the line in this exercise, we might use the vector $\langle 3, 2 \rangle$, which describes the direction of the line. Explain why the terminal points of the vectors $\mathbf{r}(t)$, where
$$\mathbf{r}(t) = \langle 2, 1 \rangle + \langle 3, 2 \rangle t,$$
trace out the graph of the line through the point $(2, -1)$ with slope $\frac{2}{3}$.

(e) Now we extend this vector approach to $\mathbb{R}^3$ and consider a second example. Let $\mathcal{L}$ be the line in $\mathbb{R}^3$ through the point $(1, 0, 2)$ in the direction of the vector $\langle 2, -1, 4 \rangle$.

Find the coordinates of three distinct points on line $\mathcal{L}$. Explain your thinking.

(f) Find a vector in form
$$\mathbf{r}(t) = \langle x_0, y_0, z_0 \rangle + \langle a, b, c \rangle t$$
whose terminal points trace out the line $\mathcal{L}$ that is described in (e). That is, you should be able to locate any point on the line by determining a corresponding value of t.

think of point slope form! mickies!

LINE

Lines in Space

In two-dimensional space, a non-vertical line is defined to be the set of points satisfying the equation
$$y = mx + b,$$
for some constants m and b. The value of m (the slope) tells us how the dependent variable changes for every one unit increase in the independent variable, while the point $(0, b)$ is the y-intercept and anchors the line to a location on the y-axis. Alternatively, we can think of the slope as being related to the vector $\langle 1, m \rangle$, which tells us the direction of the line, as shown on the left in Figure 9.49. Thus, we can identify a line in space by fixing a point P and a direction $\mathbf{v}$, as shown on the right.

Definition 9.7. A **line** in space is the set of terminal points of vectors emanating from a given point P that are parallel to a fixed vector $\mathbf{v}$.

The fixed vector $\mathbf{v}$ in the definition is called a *direction vector* for the line. As we saw in Preview Activity 9.5, to find an equation for a line through point P in the direction of vector $\mathbf{v}$, observe that

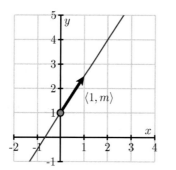 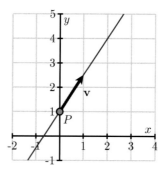

Figure 9.49: A vector description of a line

any vector parallel to **v** will have the form $t\mathbf{v}$ for some scalar t. So, any vector emanating from the point P in a direction parallel to the vector **v** will be of the form

$$\overrightarrow{OP} + \mathbf{v}t \tag{9.7}$$

for some scalar t (where O is the origin).

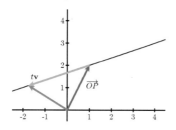 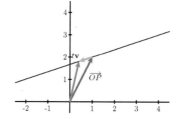 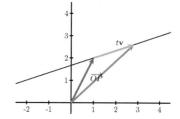

Figure 9.50: A line in 2-space.

Figure 9.50 presents three images of a line in two-space in which we can identify the vector $\overrightarrow{OP}$ and the vector $t\mathbf{v}$ as in Equation (9.7). Here, $\overrightarrow{OP}$ is the fixed vector shown in blue, while the direction vector **v** is the vector parallel to the vector shown in green (that is, the green vector represents $t\mathbf{v}$, and the line is traced out by the terminal points of the magenta vector). In other words, the tips (terminal points) of the magenta vectors (the vectors of the form $\overrightarrow{OP} + t\mathbf{v}$) trace out the line as t changes.

In particular, the terminal points of the vectors of the form in (9.7) define a linear function **r** in space of the following form, which is valid in any dimension.

The **vector form** of a line through the point P in the direction of the vector **v** is

$$\mathbf{r}(t) = \mathbf{r}_0 + t\mathbf{v}, \tag{9.8}$$

where $\mathbf{r}_0$ is the vector $\overrightarrow{OP}$ from the origin to the point P.

9.5. LINES AND PLANES IN SPACE

Of course, it is common to represent lines in the plane using the slope-intercept equation $y = mx + b$. The vector form of the line, described above, is an alternative way to represent lines that has the following two advantages. First, in two dimensions, we are able to represent vertical lines, whose slope m is not defined, using a vertical direction vector, such as $\mathbf{v} = \langle 0, 1 \rangle$. Second, this description of lines works in any dimension even though there is no concept of the slope of a line in more than two dimensions.

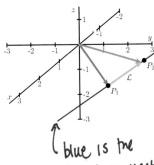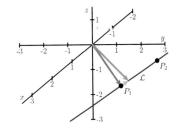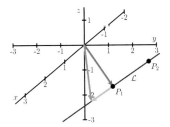

blue is the direction vector

Figure 9.51: A line in 3-space.

Activity 9.19.

Let $P_1 = (1, 2, -1)$ and $P_2 = (-2, 1, -2)$. Let $\mathcal{L}$ be the line in $\mathbb{R}^3$ through P_1 and P_2, and note that three snapshots of this line are shown in Figure 9.51.

(a) Find a direction vector for the line $\mathcal{L}$.

(b) Find a vector equation of $\mathcal{L}$ in the form $\mathbf{r}(t) = \mathbf{r}_0 + t\mathbf{v}$.

(c) Consider the vector equation $\mathbf{s}(t) = \langle -5, 0, -3 \rangle + t \langle 6, 2, 2 \rangle$. What is the direction of the line given by $\mathbf{s}(t)$? Is this new line parallel to line $\mathcal{L}$?

(d) Do $\mathbf{r}(t)$ and $\mathbf{s}(t)$ represent the same line, $\mathcal{L}$? Explain.

The Parametric Equations of a Line

The vector form of a line, $\mathbf{r}(t) = \mathbf{r}_0 + t\mathbf{v}$ in Equation (9.8), describes a line as the set of terminal points of the vectors $\mathbf{r}(t)$. If we write this in terms of components letting

$$\mathbf{r}(t) = \langle x(t), y(t), z(t) \rangle, \quad \mathbf{r}_0 = \langle x_0, y_0, z_0 \rangle, \quad \text{and} \quad \mathbf{v} = \langle a, b, c \rangle,$$

then we can equate the components on both sides of $\mathbf{r}(t) = \mathbf{r}_0 + t\mathbf{v}$ to obtain the equations

$$x(t) = x_0 + at, \quad y(t) = y_0 + bt, \quad \text{and} \quad z(t) = z_0 + ct,$$

sub set of equations defining parametric

which describe the coordinates of the points on the line. The variable t represents an arbitrary scalar and is called a *parameter*. In particular, we use the following language.

> The **parametric equations** for a line through the point $P = (x_0, y_0, z_0)$ in the direction of the vector $\mathbf{v} = \langle a, b, c \rangle$ are
> $$x(t) = x_0 + at, \quad y(t) = y_0 + bt, \quad z(t) = z_0 + ct.$$

Notice that there are many different parametric equations for the same line. For example, choosing another point P on the line or another direction vector $\mathbf{v}$ produces another set of parametric equations. It is sometimes useful to think of t as a time parameter and the parametric equations as telling us where we are on the line at each time. In this way, the parametric equations describe a particular walk taken along the line; there are, of course, many possible ways to walk along a line.

Activity 9.20.

Let $P_1 = (1, 2, -1)$ and $P_2 = (-2, 1, -2)$, and let $\mathcal{L}$ be the line in $\mathbb{R}^3$ through P_1 and P_2, which is the same line as in Activity 9.19.

(a) Find parametric equations of the line $\mathcal{L}$.

(b) Does the point $(1, 2, 1)$ lie on $\mathcal{L}$? If so, what value of t results in this point?

(c) Consider another line, $\mathcal{K}$, whose parametric equations are
$$x(s) = -2 + 4s, \quad y(s) = 1 - 3s, \quad z(s) = -2 + 2s.$$
What is the direction of line $\mathcal{K}$? $= \langle -2, 1, -2 \rangle + s \langle 4, -3, 2 \rangle$

(d) Do lines $\mathcal{L}$ and $\mathcal{K}$ intersect? If so, provide the point of intersection and the t and s values, respectively, that result in the point. If not, explain why.

direction vectors will be different!!

Planes in Space

Now that we have a way of describing lines, we would like to develop a means of describing planes in three dimensions. We studied the coordinate planes and planes parallel to them in Section 9.1. Each of those planes had one of the variables x, y, or z equal to a constant. We can note that any vector in a plane with x constant is orthogonal to the vector $\langle 1, 0, 0 \rangle$, any vector in a plane with y constant is orthogonal to the vector $\langle 0, 1, 0 \rangle$, and any vector in a plane with z constant is orthogonal to the vector $\langle 0, 0, 1 \rangle$. This idea works in general to define a plane.

> **Definition 9.8.** A **plane** p in space is the set of all terminal points of vectors emanating from a given point P_0 perpendicular to a fixed vector $\mathbf{n}$, as shown in Figure 9.52.

9.5. LINES AND PLANES IN SPACE

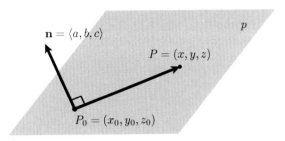

Figure 9.52: A point P_0 on a plane p with a normal vector **n**

The definition allows us to find the equation of a plane. Assume that $\mathbf{n} = \langle a, b, c \rangle$, $P_0 = (x_0, y_0, z_0)$, and that $P = (x, y, z)$ is an arbitrary point on the plane. Since the vector $\overrightarrow{PP_0}$ lies in the plane, it must be perpendicular to **n**. This means that

$$\begin{aligned}
0 &= \mathbf{n} \cdot \overrightarrow{PP_0} \\
&= \mathbf{n} \cdot \big(\langle x, y, z \rangle - \langle x_0, y_0, z_0 \rangle\big) \\
&= \mathbf{n} \cdot \langle x - x_0, y - y_0, z - z_0 \rangle \\
&= a(x - x_0) + b(y - y_0) + c(z - z_0).
\end{aligned}$$

The fixed vector **n** perpendicular to the plane is frequently called a *normal vector* to the plane. We may now summarize as follows.

> The **scalar equation** of the plane with normal vector $\mathbf{n} = \langle a, b, c \rangle$ containing the point $P_0 = (x_0, y_0, z_0)$ is
> $$a(x - x_0) + b(y - y_0) + c(z - z_0) = 0. \tag{9.9}$$

We may take this a little further and note that since

$$a(x - x_0) + b(y - y_0) + c(z - z_0) = 0,$$

it equivalently follows that

$$ax + by + cz = ax_0 + by_0 + cz_0.$$

That is, we may write the equation of a plane as $ax + by + cz = d$ where where $d = \mathbf{n} \cdot \langle x_0, y_0, z_0 \rangle$.

For instance, if we would like to describe the plane passing through the point $P_0 = (4, -2, 1)$ and perpendicular to the vector $\mathbf{n} = \langle 1, 2, 1 \rangle$, we have

$$\langle 1, 2, 1 \rangle \cdot \langle x, y, z \rangle = \langle 1, 2, 1 \rangle \cdot \langle 4, -2, 1 \rangle$$

or
$$x + 2y + z = 1.$$

Notice that the coefficients of x, y, and z in this description give a vector perpendicular to the plane. For instance, if we are presented with the plane
$$-2x + y - 3z = 4,$$
we know that $\mathbf{n} = \langle -2, 1, -3 \rangle$ is a vector perpendicular to the plane.

Activity 9.21.

(a) Write the equation of the plane p_1 passing through the point $(0, 2, 4)$ and perpendicular to the vector $\mathbf{n} = \langle 2, -1, 1 \rangle$.

(b) Is the point $(2, 0, 2)$ on the plane p_1?

(c) Write the equation of the plane p_2 that is parallel to p_1 and passing through the point $(3, 0, 4)$.

(d) Write the parametric description of the line l passing through the point $(2, 0, 2)$ and perpendicular to the plane p_3 described the equation $x + 2y - 2z = 7$.

(e) Find the point at which l intersects the plane p_3.

◁

Just as two distinct points in space determine a line, three non-collinear points in space determine a plane. Consider three points P_0, P_1, and P_2 in space, not all lying on the same line as shown in Figure 9.53.

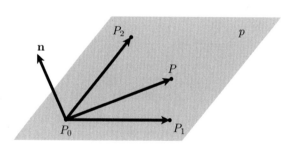

Figure 9.53: A plane determined by three points P_0, P_1, and P_2

Observe that the vectors $\overrightarrow{P_0P_1}$ and $\overrightarrow{P_0P_2}$ both lie in the plane p. If we form their cross-product
$$\mathbf{n} = \overrightarrow{P_0P_1} \times \overrightarrow{P_0P_2},$$
we obtain a normal vector to the plane p. Therefore, if P is any other point on p, it then follows that $\overrightarrow{P_0P}$ will be perpendicular to $\mathbf{n}$, and we have, as before, the equation
$$\mathbf{n} \cdot \overrightarrow{P_0P} = 0. \tag{9.10}$$

9.5. LINES AND PLANES IN SPACE

Activity 9.22.

Let $P_0 = (1, 2, -1)$, $P_1 = (1, 0, -1)$, and $P_2 = (0, 1, 3)$ and let p be the plane containing P_0, P_1, and P_2.

(a) Determine the components of the vectors $\overrightarrow{P_0P_1}$ and $\overrightarrow{P_0P_2}$.

(b) Find a normal vector **n** to the plane p.

(c) Find the scalar equation of the plane p.

(d) Consider a second plane, q, whose scalar equation is $-3(x-1) + 4(y+3) + 2(z-5) = 0$. Find two different points on plane q, as well as a vector **m** that is normal to q.

(e) We define the angle between two planes to be the angle between their respective normal vectors. What is the angle between planes p and q?

◁

Summary

- While lines in $\mathbb{R}^3$ do not have a slope, like lines in $\mathbb{R}^2$ they can be characterized by a point and a direction vector. Indeed, we define a line in space to be the set of terminal points of vectors emanating from a given point that are parallel to a fixed vector.

- Vectors play a critical role in representing the equation of a line. In particular, the terminal points of the vector $\mathbf{r}(t) = \mathbf{r}_0 + t\mathbf{v}$ define a linear function $\mathbf{r}$ in space through the terminal point of the vector $\mathbf{r}_0$ in the direction of the vector $\mathbf{v}$, tracing out a line in space.

- A plane in space is the set of all terminal points of vectors emanating from a given point perpendicular to a fixed vector.

- If P_1, P_2, and P_3 are non-collinear points in space, the vectors $\overrightarrow{P_1P_2}$ and and $\overrightarrow{P_1P_3}$ are vectors in the plane and the vector $\mathbf{n} = \overrightarrow{P_1P_2} \times \overrightarrow{P_1P_3}$ is a normal vector to the plane. So any point P in the plane satisfies the equation $\overrightarrow{PP_1} \cdot \mathbf{n} = 0$. If we let $P = (x, y, z)$, $\mathbf{n} = \langle a, b, c \rangle$ be the normal vector, and $P_1 = (x_0, y_0, z_0)$, we can also represent the plane with the equation

$$a(x - x_0) + b(y - y_0) + c(z - z_0) = 0.$$

Exercises

1. The vector and parametric forms of a line allow us to easily describe line segments in space.

   Let $P_1 = (1, 2, -1)$ and $P_2 = (-2, 1, -2)$, and let $\mathcal{L}$ be the line in $\mathbb{R}^3$ through P_1 and P_2 as in Activity 9.19.

   (a) What value of the parameter t makes $(x(t), y(t), z(t)) = P_1$? What value of t makes $(x(t), y(t), z(t)) = P_2$? *you are trying to restrict t's length*

   (b) What restrictions on the parameter t describe the line segment between the points P_1 and P_2? *not the same vector!*

(c) What about the line segment (along the same line) from $(7, 4, 1)$ to $(-8, -1, -4)$?

(d) Now, consider a segment that lies on a different line: parameterize the segment that connects point $R = (4, -2, 7)$ to $Q = (-11, 4, 27)$ in such a way that $t = 0$ corresponds to point Q, while $t = 2$ corresponds to R.

2. This exercise explores key relationships between a pair of lines. Consider the following two lines: one with parametric equations $x(s) = 4 - 2s$, $y(s) = -2 + s$, $z(s) = 1 + 3s$, and the other being the line through $(-4, 2, 17)$ in the direction $\mathbf{v} = \langle -2, 1, 5 \rangle$.

   (a) Find a direction vector for the first line, which is given in parametric form.

   (b) Find parametric equations for the second line, written in terms of the parameter t.

   (c) Show that the two lines intersect at a single point by finding the values of s and t that result in the same point.

   (d) Find the angle formed where the two lines intersect, noting that this angle will be given by the angle between their respective direction vectors.

   (e) Find an equation for the plane that contains both of the lines described in this problem.

3. This exercise explores key relationships between a pair of planes. Consider the following two planes: one with scalar equation $4x - 5y + z = -2$, and the other which passes through the points $(1, 1, 1)$, $(0, 1, -1)$, and $(4, 2, -1)$.

   (a) Find a vector normal to the first plane.

   (b) Find the scalar equation for the second plane.

   (c) Find the angle between the planes, where the angle between them is defined by the angle between their respective normal vectors.

   (d) Find a point that lies on both planes.

   (e) Since these two planes do not have parallel normal vectors, the planes must intersect, and thus must intersect in a line. Observe that the line of intersection lies in both planes, and thus the direction vector of the line must be perpendicular to each of the respective normal vectors of the two planes. Find a direction vector for the line of intersection for the two planes.

   (f) Determine parametric equations for the line of intersection of the two planes.

4. In this problem, we explore how we can use what we know about vectors and projections to find the distance from a point to a plane.

   Let p be the plane with equation $z = -4x + 3y + 4$, and let $Q = (4, -1, 8)$.

   (a) Show that Q does not lie in the plane p.

   (b) Find a normal vector $\mathbf{n}$ to the plane p.

9.5. LINES AND PLANES IN SPACE

(c) Find the coordinates of a point P in p.

(d) Find the components of $\overrightarrow{PQ}$. Draw a picture to illustrate the objects found so far.

(e) Explain why $|\text{comp}_n \overrightarrow{PQ}|$ gives the distance from the point Q to the plane p. Find this distance.
   ↑ abs value signs. distance can't be negative.

9.6 Vector-Valued Functions

> **Motivating Questions**
>
> *In this section, we strive to understand the ideas generated by the following important questions:*
>
> - What is a vector-valued function? What do we mean by the graph of a vector-valued function?
>
> - What is a parameterization of a curve in $\mathbb{R}^2$? In $\mathbb{R}^3$? What can the parameterization of a curve can tell us?

Introduction

So far, we have seen several different examples of curves in space, including traces and contours of functions of two variables, as well as lines in 3-space. Recall that for a line through a fixed point $\mathbf{r}_0$ in the direction of vector $\mathbf{v}$, we may express the line parametrically through the single vector equation

$$\mathbf{r}(t) = \mathbf{r}_0 + t\mathbf{v}.$$

From this perspective, the vector $\mathbf{r}(t)$ is a function that depends on the parameter t, and the terminal points of this vector trace out the line in space.

Like lines, other curves in space are one-dimensional objects, and thus we aspire to similarly express the coordinates of points on a given curve in terms of a single variable. Vectors are a perfect vehicle for doing so – we can use vectors based at the origin to identify points in space, and connect the terminal points of these vectors to draw a curve in space. This approach will allow us to draw an incredible variety of graphs in 2- and 3-space, as well as to identify and describe curves in n-space for any n.

Preview Activity 9.6. In this activity we consider how we might use vectors to define a curve in space.

(a) On a single set of axes in $\mathbb{R}^2$, draw the vectors $\langle \cos(0), \sin(0) \rangle$, $\langle \cos\left(\frac{\pi}{2}\right), \sin\left(\frac{\pi}{2}\right) \rangle$, $\langle \cos(\pi), \sin(\pi) \rangle$, and $\langle \cos\left(\frac{3\pi}{2}\right), \sin\left(\frac{3\pi}{2}\right) \rangle$ with their initial points at the origin.

(b) On the same set of axes, draw the vectors $\langle \cos\left(\frac{\pi}{4}\right), \sin\left(\frac{\pi}{4}\right) \rangle$, $\langle \cos\left(\frac{3\pi}{4}\right), \sin\left(\frac{3\pi}{4}\right) \rangle$, $\langle \cos\left(\frac{5\pi}{4}\right), \sin\left(\frac{5\pi}{4}\right) \rangle$, and $\langle \cos\left(\frac{7\pi}{4}\right), \sin\left(\frac{7\pi}{4}\right) \rangle$ with their initial points at the origin.

(c) Based on the pictures from parts (a) and (b), sketch the set of *terminal* points of all of the vectors of the form $\langle \cos(t), \sin(t) \rangle$, where t assumes values from 0 to 2π. What is the resulting figure? Why?

Vector-Valued Functions

Consider the curve shown in Figure 9.54. As in Preview Activity 9.6, we can think of a point on this curve as resulting from a vector from the origin to the point. As the point travels along the curve, the vector changes in order to terminate at the desired point. A few still pictures of this motion are shown in Figure 9.54.

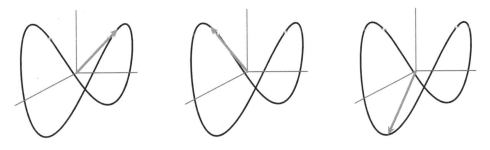

Figure 9.54: The graph of a curve in space.

Thus, we can think of the curve as a collection of terminal points of vectors emanating from the origin. We therefore view a point traveling along this curve as a function of time t, and define a function $\mathbf{r}$ whose input is the variable t and whose output is the vector from the origin to the point on the curve at time t. In so doing, we have introduced a new type of function, one whose input is a scalar and whose output is a vector.

The terminal points of the vector outputs of $\mathbf{r}$ then trace out the curve in space. From this perspective, the x, y, and z coordinates of the point are functions of time, t, say

$$x = x(t), \quad y = y(t), \quad \text{and} \quad z = z(t),$$

and thus we have three coordinate functions that enable us to represent the curve. The variable t is called a *parameter* and the equations $x = x(t)$, $y = y(t)$, and $z = z(t)$ are called *parametric equations* (or a *parameterization of the curve*). The function $\mathbf{r}$ whose output is the vector from the origin to a point on the curve is defined by

$$\mathbf{r}(t) = \langle x(t), y(t), z(t) \rangle.$$

Note that the input of $\mathbf{r}$ is the real-valued parameter t and the corresponding output is vector $\langle x(t), y(t), z(t) \rangle$. Such a function is called a *vector-valued function* because each real number input generates a vector output. More formally, we state the following definition.

> **Definition 9.9.** A **vector-valued function** is a function whose input is a real parameter t and whose output is a vector that depends on t. The **graph** of a vector-valued function is the set of all terminal points of the output vectors with their initial points at the origin.
>
> **Parametric equations** for a curve are equations of the form
> $$x = x(t), \quad y = y(t), \quad \text{and} \quad z = z(t)$$
> that describe the (x, y, z) coordinates of a point on a curve in $\mathbb{R}^3$.

Note particularly that every set of parametric equations determines a vector-valued function of the form
$$\mathbf{r}(t) = \langle x(t), y(t), z(t) \rangle,$$
and every vector-valued function defines a set of parametric equations for a curve. Moreover, we can consider vector-valued functions and parameterizations in $\mathbb{R}^2$, $\mathbb{R}^4$, or indeed a real space of any dimension. As a reminder, in Section 9.5, we determined the parametric equations of a line in space using a point and a direction vector. For a nonlinear example, the curve in Figure 9.54 has the parametric equations
$$x(t) = \cos(t), \quad y(t) = \sin(t), \quad \text{and} \quad z(t) = \cos(t)\sin(t).$$

Represented as a vector-valued function $\mathbf{r}$, the curve in Figure 9.54 is the graph of
$$\mathbf{r}(t) = \langle \cos(t), \sin(t), \cos(t)\sin(t) \rangle.$$

Activity 9.23.

The same curve can be represented with different parameterizations. Use your calculator,[8] Wolfram|Alpha, or some other graphing device[9] to plot the curves generated by the following vector-valued functions. Compare and contrast the graphs – explain how they are alike and how they are different.

(a) $\mathbf{r}(t) = \langle \sin(t), \cos(t) \rangle$

(b) $\mathbf{r}(t) = \langle \sin(2t), \cos(2t) \rangle$

(c) $\mathbf{r}(t) = \langle \cos(t + \pi), \sin(t + \pi) \rangle$

◁

---

[8] If you have a graphing calculator you can draw graphs of vector-valued functions in $\mathbb{R}^2$ using the parametric mode (often found in the MODE menu).

[9] e.g., http://webspace.ship.edu/msrenault/ggb/parametric_grapher.html

9.6. VECTOR-VALUED FUNCTIONS

The examples in Activity 9.23 illustrate that a parameterization allows us to look not only at the graph, but at the direction and speed at which the graph is traversed as t changes. In the different parameterizations of the circle, we see that we can start at different points and move around the circle in either direction. The calculus of vector-valued functions – which we will begin to investigate in Section 9.7 – will enable us to precisely quantify the direction, speed, and acceleration of a particle moving along a curve in space. As such, describing curves parametrically will allow us to not only indicate the curve itself, but also to describe how motion occurs along the curve.

Using parametric equations to define vector-valued functions in two dimensions is much more versatile than just defining y as a function of x. In fact, if $y = f(x)$ is a function of x, then we can parameterize the graph of f by

$$\mathbf{r}(t) = \langle t, f(t) \rangle,$$

and thus every single-variable function may be described parametrically. In addition, as we saw in Preview Activity 9.6 and Activity 9.23, we can use vector-valued functions to represent curves in the plane that do not define y as a function of x (or x as a function of y).[10]

Activity 9.24.

Vector-valued functions can be used to generate many interesting curves. Graph each of the following using an appropriate tool[11], and then write one sentence for each function to describe the behavior of the resulting curve.

(a) $\mathbf{r}(t) = \langle t\cos(t), t\sin(t) \rangle$

(b) $\mathbf{r}(t) = \langle \sin(t)\cos(t), t\sin(t) \rangle$

(c) $\mathbf{r}(t) = \langle t^2 \sin(t)\cos(t), 0.9t\cos(t^2), \sin(t) \rangle$

(d) $\mathbf{r}(t) = \langle \sin(5t), \sin(4t) \rangle$

(e) Experiment with different formulas for $x(t)$ and $y(t)$ and ranges for t to see what other interesting curves you can generate. Share your best results with peers.

◁

Recall from our earlier work that the traces and level curves of a function are themselves curves in space. Thus, we may determine parameterizations for them. For example, if $z = f(x,y) = \cos(x^2+y^2)$, the $y=1$ trace of the function is given by setting $y=1$ and letting x be parameterized by the variable t; then, the trace is the curve whose parameterization is $\langle t, 1, \cos(t^2+1) \rangle$.

Activity 9.25.

Consider the paraboloid defined by $f(x,y) = x^2 + y^2$.

---

[10] As an aside, vector-valued functions make it easy to plot the inverse of a one-to-one function in two dimensions. To see how, if $y = f(x)$ defines a one-to-one function, then we can parameterize this function by $\mathbf{r}(t) = \langle t, f(t) \rangle$. Since the inverse function just reverses the role of input and output, a parameterization for f^{-1} is $\langle f(t), t \rangle$.

[11] e.g., the 2D grapher at http://webspace.ship.edu/msrenault/ggb/parametric_grapher.html, or for 3D graphs Wolfram|Alpha, an on-line 3D grapher like http://www.math.uri.edu/~bkaskosz/flashmo/parcur/, or some other device

(a) Find a parameterization for the $x = 2$ trace of f. What type of curve does this trace describe?

(b) Find a parameterization for the $y = -1$ trace of f. What type of curve does this trace describe?

(c) Find a parameterization for the level curve $f(x, y) = 25$. What type of curve does this trace describe?

(d) How do your responses change to all three of the preceding question if you instead consider the function g defined by $g(x, y) = x^2 - y^2$? (Hint for generating one of the parameterizations: $\sec^2(t) - \tan^2(t) = 1$.)

◁

Summary

- A vector-valued function is a function whose input is a real parameter t and whose output is a vector that depends on t. The graph of a vector-valued function is the set of all terminal points of the output vectors with their initial points at the origin.

- Every vector-valued function provides a parameterization of a curve. In $\mathbb{R}^2$, a parameterization of a curve is a pair of equations $x = x(t)$ and $y = y(t)$ that describes the coordinates of a point (x, y) on the curve in terms of a parameter t. In $\mathbb{R}^3$, a parameterization of a curve is a set of three equations $x = x(t)$, $y = y(t)$, and $z = z(t)$ that describes the coordinates of a point (x, y, z) on the curve in terms of a parameter t.

Exercises

1. The standard parameterization for the unit circle is $\langle \cos(t), \sin(t) \rangle$, for $0 \le t \le 2\pi$.

   (a) Find a vector-valued function **r** that describes a point traveling along the unit circle so that at time $t = 0$ the point is at $\left(\frac{\sqrt{2}}{2}, \frac{\sqrt{2}}{2}\right)$ and travels clockwise along the circle as t increases.

   (b) Find a vector-valued function **r** that describes a point traveling along the unit circle so that at time $t = 0$ the point is at $\left(\frac{\sqrt{2}}{2}, \frac{\sqrt{2}}{2}\right)$ and travels counter-clockwise along the circle as t increases.

   (c) Find a vector-valued function **r** that describes a point traveling along the unit circle so that at time $t = 0$ the point is at $\left(-\frac{\sqrt{2}}{2}, \frac{\sqrt{2}}{2}\right)$ and travels clockwise along the circle as t increases.

2. Let a and b be positive real numbers. You have probably seen the equation $\frac{(x-h)^2}{a^2} + \frac{(y-k)^2}{b^2} = 1$ that generates an ellipse, centered at (h, k), with a horizontal axis of length $2a$ and a vertical axis of length $2b$.

(a) Explain why the vector function **r** defined by $\mathbf{r}(t) = \langle a\cos(t), b\sin(t)\rangle$, $0 \le t \le 2\pi$ is one parameterization of the ellipse $\frac{x^2}{a^2} + \frac{y^2}{b^2} = 1$.

(b) Find a parameterization of the ellipse $\frac{x^2}{4} + \frac{y^2}{16} = 1$ that is traversed counterclockwise.

(c) Find a parameterization of the ellipse $\frac{(x+3)^2}{4} + \frac{(y-2)^2}{9} = 1$.

(d) Determine the x-y equation of the ellipse that is parameterized by
$$\mathbf{r}(t) = \langle 3 + 4\sin(2t), 1 + 3\cos(2t)\rangle.$$

3. Consider the two-variable function $z = f(x,y) = 3x^2 + 4y^2 - 2$.

   (a) Determine a vector-valued function **r** that parameterizes the curve which is the $x = 2$ trace of $z = f(x,y)$. Plot the resulting curve. Do likewise for $x = -2, -1, 0$, and 1.

   (b) Determine a vector-valued function **r** that parameterizes the curve which is the $y = 2$ trace of $z = f(x,y)$. Plot the resulting curve. Do likewise for $y = -2, -1, 0$, and 1.

   (c) Determine a vector-valued function **r** that parameterizes the curve which is the $z = 2$ contour of $z = f(x,y)$. Plot the resulting curve. Do likewise for $z = -2, -1, 0$, and 1.

   (d) Use the traces and contours you've just investigated to create a wireframe plot of the surface generated by $z = f(x,y)$. In addition, write two sentences to describe the characteristics of the surface.

4. Recall that any line in space may be represented parametrically by a vector-valued function.

   (a) Find a vector-valued function **r** that parameterizes the line through $(-2, 1, 4)$ in the direction of the vector $\mathbf{v} = \langle 3, 2, -5\rangle$.

   (b) Find a vector-valued function **r** that parameterizes the line of intersection of the planes $x + 2y - z = 4$ and $3x + y - 2z = 1$.

   (c) Determine the point of intersection of the lines given by
   $$x = 2 + 3t,\ y = 1 - 2t,\ z = 4t,$$
   $$x = 3 + 1s,\ y = 3 - 2s,\ z = 2s.$$
   Then, find a vector-valued function **r** that parameterizes the line that passes through the point of intersection you just found and is perpendicular to both of the given lines.

9.7 Derivatives and Integrals of Vector-Valued Functions

Motivating Questions

In this section, we strive to understand the ideas generated by the following important questions:

- What do we mean by the derivative of a vector-valued function and how do we calculate it?
- What does the derivative of a vector-valued function measure?
- What do we mean by the integral of a vector-valued function and how do we compute it?
- How do we describe the motion of a projectile if the only force acting on the object is acceleration due to gravity?

Introduction

A vector-valued function $\mathbf{r}$ determines a curve in space as the collection of terminal points of the vectors $\mathbf{r}(t)$. If the curve is smooth, it is natural to ask whether $\mathbf{r}(t)$ has a derivative. In the same way, our experiences with integrals in single-variable calculus prompt us to wonder what the integral of a vector-valued function might be and what it might tell us. We explore both of these questions in detail in this section.

For now, let's recall some important ideas from calculus I. Given a function s that measures the position of an object moving along an axis, its derivative, s', is defined by

$$s'(t) = \lim_{h \to 0} \frac{s(t+h) - s(t)}{h},$$

and measures the instantaneous rate of change of s with respect to time. In particular, for a fixed value $t = a$, $s'(a)$ measures the velocity of the moving object, as well as the slope of the tangent line to the curve $y = s(t)$ at the point $(a, s(a))$.

As we work with vector-valued functions, we will strive to update these ideas and perspectives into the context of curves in space and outputs that are vectors.

Preview Activity 9.7. Let $\mathbf{r}(t) = \cos(t)\mathbf{i} + \sin(2t)\mathbf{j}$ describe the path traveled by an object at time t.

(a) Use appropriate technology to help you sketch the graph of the vector-valued function $\mathbf{r}$, and then locate and label the point on the graph when $t = \pi$.

(b) Recall that for functions of a single variable, the derivative of a sum is the sum of the derivatives; that is, $\frac{d}{dx}[f(x) + g(x)] = f'(x) + g'(x)$. With this idea in mind and viewing $\mathbf{i}$ and $\mathbf{j}$ as constant vectors, what do you expect the derivative of $\mathbf{r}$ to be? Write a proposed formula for $\mathbf{r}'(t)$.

(c) Use your result from part (b) to compute $\mathbf{r}'(\pi)$. Sketch this vector $\mathbf{r}'(\pi)$ as emanating from the point on the graph of $\mathbf{r}$ when $t = \pi$, and explain what you think $\mathbf{r}'(\pi)$ tells us about the object's motion.

⋈

The Derivative

In single variable calculus, we define the derivative, f', of a given function f by

$$f'(x) = \lim_{h \to 0} \frac{f(x+h) - f(x)}{h},$$

provided the limit exists. At a given value of a, $f'(a)$ measures the instantaneous rate of change of f, and also tells us the slope of the tangent line to the curve $y = f(x)$ at the point $(a, f(a))$. The definition of the derivative extends naturally to vector-valued functions and curves in space.

Definition 9.10. The **derivative** of a vector-valued function $\mathbf{r}$ is defined to be

$$\mathbf{r}'(t) = \lim_{h \to 0} \frac{\mathbf{r}(t+h) - \mathbf{r}(t)}{h}$$

for those values of t at which the limit exists. We also use the notation $\frac{d\mathbf{r}}{dt}$ and $\frac{d}{dt}[\mathbf{r}(t)]$ for $\mathbf{r}'(t)$.

Activity 9.26.

Let's investigate how we can interpret the derivative $\mathbf{r}'(t)$. Let $\mathbf{r}$ be the vector-valued function whose graph is shown in Figure 9.55, and let h be a scalar that represents a small change in time. The vector $\mathbf{r}(t)$ is the blue vector in Figure 9.55 and $\mathbf{r}(t+h)$ is the green vector.

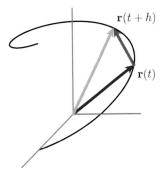

Figure 9.55: A single difference quotient.

(a) Is the quantity $\mathbf{r}(t+h) - \mathbf{r}(t)$ a vector or a scalar? Identify this object in Figure 9.55.

(b) Is $\frac{\mathbf{r}(t+h)-\mathbf{r}(t)}{h}$ a vector or a scalar? Sketch a representative vector $\frac{\mathbf{r}(t+h)-\mathbf{r}(t)}{h}$ with $h < 1$ in Figure 9.55.

(c) Think of $\mathbf{r}(t)$ as providing the position of an object moving along the curve these vectors trace out. What do you think that the vector $\frac{\mathbf{r}(t+h)-\mathbf{r}(t)}{h}$ measures? Why? (Hint: You might think analogously about difference quotients such as $\frac{f(x+h)-f(x)}{h}$ or $\frac{s(t+h)-s(t)}{h}$ from calculus I.)

(d) Figure 9.56 presents three snapshots of the vectors $\frac{\mathbf{r}(t+h)-\mathbf{r}(t)}{h}$ as we let $h \to 0$. Write 2-3 sentences to describe key attributes of the vector

$$\lim_{h \to 0} \frac{\mathbf{r}(t+h) - \mathbf{r}(t)}{h}.$$

(Hint: Compare to limits such as $\lim_{h \to 0} \frac{f(x+h)-f(x)}{h}$ or $\lim_{h \to 0} \frac{s(t+h)-s(t)}{h}$ from calculus I, keeping in mind that in three dimensions there is no general concept of slope.)

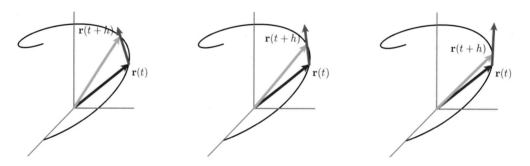

Figure 9.56: Snapshots of several difference quotients.

◁

As Activity 9.26 indicates, if $\mathbf{r}(t)$ determines the position of an object at time t, then $\frac{\mathbf{r}(t+h)-\mathbf{r}(t)}{h}$ represents the average rate of change in the position of the object over the interval $[t, t+h]$, which is also the average velocity of the object on this interval. Thus, the derivative

$$\mathbf{r}'(t) = \lim_{h \to 0} \frac{\mathbf{r}(t+h) - \mathbf{r}(t)}{h}$$

is the instantaneous rate of change of $\mathbf{r}(t)$ at time t (for those values of t for which the limit exists), so $\mathbf{r}'(t) = \mathbf{v}(t)$ is the instantaneous velocity of the object at time t. Furthermore, we can interpret the derivative $\mathbf{r}'(t)$ as the direction vector of the line tangent to the graph of $\mathbf{r}$ at the value t.

Similarly,

$$\mathbf{v}'(t) = \mathbf{r}''(t) = \lim_{h \to 0} \frac{\mathbf{v}(t+h) - \mathbf{v}(t)}{h}$$

is the instantaneous rate of change of the velocity of the object at time t, for those values of t for which the limits exists, and thus $\mathbf{v}'(t) = \mathbf{a}(t)$ is the acceleration of the moving object.

9.7. DERIVATIVES AND INTEGRALS OF VECTOR-VALUED FUNCTIONS

Note well: Both the velocity and acceleration are *vector quantities*: they have magnitude and direction. By contrast, the magnitude of the velocity vector, $|\mathbf{v}(t)|$, which is the *speed* of the object at time t, is a scalar quantity.

Computing Derivatives

As we learned in single variable calculus, computing derivatives from the definition is often difficult. Fortunately, properties of the limit make it straightforward to calculate the derivative of a vector-valued function similar to how we developed shortcut differentiation rules in calculus I. To see why, recall that the limit of a sum is the sum of the limits, and that we can remove constant factors from limits. Thus, as we observed in a particular example in Preview Activity 9.7, if $\mathbf{r}(t) = x(t)\mathbf{i} + y(t)\mathbf{j} + z(t)\mathbf{k}$, it follows that

$$\mathbf{r}'(t) = \lim_{h \to 0} \frac{\mathbf{r}(t+h) - \mathbf{r}(t)}{h}$$
$$= \lim_{h \to 0} \frac{[x(t+h) - x(t)]\mathbf{i} + [y(t+h) - y(t)]\mathbf{j} + [z(t+h) - z(t)]\mathbf{k}}{h}$$
$$= \left(\lim_{h \to 0} \frac{x(t+h) - x(t)}{h}\right)\mathbf{i} + \left(\lim_{h \to 0} \frac{y(t+h) - y(t)}{h}\right)\mathbf{j} + \left(\lim_{h \to 0} \frac{z(t+h) - z(t)}{h}\right)\mathbf{k}$$
$$= x'(t)\mathbf{i} + y'(t)\mathbf{j} + z'(t)\mathbf{k}.$$

Thus, we can calculate the derivative of a vector-valued function by simply differentiating its components.

If $\mathbf{r}(t) = x(t)\mathbf{i} + y(t)\mathbf{j} + z(t)\mathbf{k}$, then

$$\frac{d}{dt}\mathbf{r}(t) = x'(t)\mathbf{i} + y'(t)\mathbf{j} + z'(t)\mathbf{k}$$

for those values of t at which x, y, and z are differentiable.

Activity 9.27. ☑ HW!!

For each of the following vector-valued functions, find $\mathbf{r}'(t)$.

(a) $\mathbf{r}(t) = \langle \cos(t), t\sin(t), \ln(t) \rangle$. ✓

(b) $\mathbf{r}(t) = \langle t^2 + 3t, e^{-2t}, \frac{t}{t^2+1} \rangle$. ✓

(c) $\mathbf{r}(t) = \langle \tan(t), \cos(t^2), te^{-t} \rangle$. ✓

(d) $\mathbf{r}(t) = \langle \sqrt{t^4 + 4}, \sin(3t), \cos(4t) \rangle$. ✓

◁

In first-semester calculus, we developed several important differentiation rules, including the constant multiple, product, quotient, and chain rules. For instance, recall that we formally state

9.7. DERIVATIVES AND INTEGRALS OF VECTOR-VALUED FUNCTIONS

the product rule as

$$\frac{d}{dx}[f(x) \cdot g(x)] = f(x) \cdot g'(x) + g(x) \cdot f'(x).$$

There are several analogous rules for vector-valued functions, including a product rule for scalar functions and vector-valued functions. These rules, which are easily verified, are summarized in the following theorem.

Theorem. Let f be a differentiable real-valued function of a real variable t and let **r** and **s** be differentiable vector-valued functions of the real variable t. Then

1. $\dfrac{d}{dt}[\mathbf{r}(t) + \mathbf{s}(t)] = \mathbf{r}'(t) + \mathbf{s}'(t)$ sum of $r'(t) = r_1'(t) + r_2'(t)$

2. $\dfrac{d}{dt}[f(t)\mathbf{r}(t)] = f(t)\mathbf{r}'(t) + f'(t)\mathbf{r}(t)$ product rule
   gives # ↑ gives vector

3. $\dfrac{d}{dt}[\mathbf{r}(t) \cdot \mathbf{s}(t)] = \mathbf{r}'(t) \cdot \mathbf{s}(t) + \mathbf{r}(t) \cdot \mathbf{s}'(t)$ product rule again?

4. $\dfrac{d}{dt}[\mathbf{r}(t) \times \mathbf{s}(t)] = \mathbf{r}'(t) \times \mathbf{s}(t) + \mathbf{r}(t) \times \mathbf{s}'(t)$ cross product

5. $\dfrac{d}{dt}[\mathbf{r}(f(t))] = f'(t)\mathbf{r}'(f(t))$. chain rule

Note well. When applying these properties, use care to interpret the quantities involved as either scalars or vectors. For example, $\mathbf{r}(t) \cdot \mathbf{s}(t)$ defines a scalar function because we have taken the dot product of two vector-valued functions. However, $\mathbf{r}(t) \times \mathbf{s}(t)$ defines a vector-valued function since we have taken the cross product of two vector-valued functions.

Activity 9.28. ☑ HW!!

The left side of Figure 9.57 shows the curve described by the vector-valued function **r** defined by

$$\mathbf{r}(t) = \left\langle 2t - \frac{1}{2}t^2 + 1, t - 1 \right\rangle.$$

(a) Find the object's velocity $\mathbf{v}(t)$. 1st

(b) Find the object's acceleration $\mathbf{a}(t)$. second

(c) Indicate on the left of Figure 9.57 the object's position, velocity and acceleration at the times $t = 0, 2, 4$. Draw the velocity and acceleration vectors with their tails placed at the object's position.

(d) Recall that the speed is $|\mathbf{v}| = \sqrt{\mathbf{v} \cdot \mathbf{v}}$. Find the object's speed and graph it as a function of time t on the right of Figure 9.57. When is the object's speed the slowest? When is the speed increasing? When it is decreasing?

9.7. DERIVATIVES AND INTEGRALS OF VECTOR-VALUED FUNCTIONS

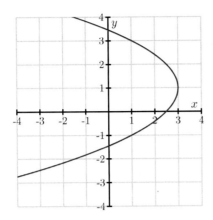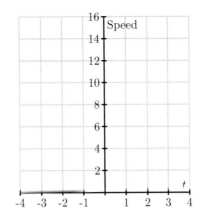

Figure 9.57: The curve $\mathbf{r}(t) = \langle 2t - \frac{1}{2}t^2 + 1, t - 1 \rangle$ and its speed.

(e) What seems to be true about the angle between **v** and **a** when the speed is at a minimum? What is the angle between **v** and **a** when the speed is increasing? when the speed is decreasing?

(f) Since the square root is an increasing function, we see that the speed increases precisely when $\mathbf{v} \cdot \mathbf{v}$ is increasing. Use the product rule for the dot product to express $\frac{d}{dt}(\mathbf{v} \cdot \mathbf{v})$ in terms of the velocity **v** and acceleration **a**. Use this to explain why the speed is increasing when $\mathbf{v} \cdot \mathbf{a} > 0$ and decreasing when $\mathbf{v} \cdot \mathbf{a} < 0$. Compare this to part (d).

(g) Show that the speed's rate of change is
$$\frac{d}{dt}|\mathbf{v}(t)| = \text{comp}_{\mathbf{v}} \mathbf{a}.$$

◁

Tangent Lines

One of the most important ideas in first-semester calculus is that a differentiable function is *locally linear*: that is, when viewed up close, the curve generated by a differentiable function looks very much like a line. Indeed, when we zoom in sufficiently far on a particular point, the curve looks indistinguishable from its tangent line.

In the same way, we expect that a smooth curve in 3-space will be locally linear. In the following activity, we investigate how to find the tangent line to such a curve. Recall from our work in Section 9.5 that the vector equation of a line that passes through the point at the tip of the vector $\mathbf{L}_0 = \langle x_0, y_0, z_0 \rangle$ in the direction of the vector $\mathbf{u} = \langle a, b, c \rangle$ can be written as
$$\mathbf{L}(t) = \mathbf{L}_0 + t\mathbf{u}.$$

In parametric form, the line **L** is given by
$$x(t) = x_0 + at, \quad y(t) = y_0 + bt, \quad z(t) = z_0 + ct.$$

9.7. DERIVATIVES AND INTEGRALS OF VECTOR-VALUED FUNCTIONS

Activity 9.29.

Let
$$\mathbf{r}(t) = \cos(t)\mathbf{i} - \sin(t)\mathbf{j} + t\mathbf{k}.^{12}$$

(a) Determine the coordinates of the point on the curve traced out by $\mathbf{r}(t)$ when $t = \pi$.

(b) Find a direction vector for the line tangent to the graph of $\mathbf{r}$ at the point where $t = \pi$.

(c) Find the parametric equations of the line tangent to the graph of $\mathbf{r}$ when $t = \pi$.

(d) Sketch a plot of the curve $\mathbf{r}(t)$ and its tangent line near the point where $t = \pi$. In addition, include a sketch of $\mathbf{r}'(\pi)$. What is the important role of $\mathbf{r}'(\pi)$ in this activity?

◁

We see that our work in Activity 9.29 can be generalized. Given a differentiable vector-valued function $\mathbf{r}$, the tangent line to the curve at the input value a is given by

$$\mathbf{L}(t) = \mathbf{r}(a) + t\mathbf{r}'(a). \tag{9.11}$$

Here we see that because the tangent line is determined entirely by a given point and direction, the point is provided by the function $\mathbf{r}$, evaluated at $t = a$, while the direction is provided by the derivative, $\mathbf{r}'$, again evaluated at $t = a$. Note how analogous the formula for $\mathbf{L}(t)$ is to the tangent line approximation from single-variable calculus: in that context, for a given function $y = f(x)$ at a value $x = a$, we found that the tangent line can be expressed by the linear function $y = L(x)$ whose formula is

$$L(x) = f(a) + f'(a)(x - a).$$

Equation (9.11) for the tangent line $\mathbf{L}(t)$ to the vector-valued function $\mathbf{r}(t)$ is nearly identical. Indeed, because there are multiple parameterizations for a single line, it is even possible[13] to write the parameterization as

$$\mathbf{L}(t) = \mathbf{r}(a) + (t - a)\mathbf{r}'(a). \tag{9.12}$$

As we will learn more in Chapter 10, a smooth surface in 3-space is also locally linear. That means that the surface will look like a plane, which we call its *tangent plane*, as we zoom in on the graph. It is possible to use tangent lines to traces of the surface to generate a formula for the tangent plane; see Exercise 3 at the end of this section for more details.

Integrating a Vector-Valued Function

Recall from single variable calculus that an antiderivative of a function f of the independent variable x is a function F that satisfies $F'(x) = f(x)$. We then defined the indefinite integral $\int f(x)\, dx$

---

[12]You can sketch the graph with Wolfram Alpha, the applet at http://gvsu.edu/s/LR, or some other appropriate technology.

[13]In Equation (9.11), $\mathbf{L}(0) = \mathbf{r}(a)$, so the line's parameterization "starts" at $t = 0$. When we write the parameterization in the form of Equation (9.12), $\mathbf{L}(a) = \mathbf{r}(a)$, so the line's parameterization "starts" at $t = a$.

9.7. DERIVATIVES AND INTEGRALS OF VECTOR-VALUED FUNCTIONS

to be the general antiderivative of f. Recall that the general antiderivative includes an added constant C in order to indicate that the general antiderivative is in fact an entire family of functions. We can do the similar work with vector-valued functions.

> **Definition 9.11.** An **antiderivative** of a vector-valued function $\mathbf{r}$ is a vector-valued function $\mathbf{R}$ such that
> $$\mathbf{R}'(t) = \mathbf{r}(t).$$
> The **indefinite integral** $\int \mathbf{r}(t)\,dt$ of a vector-valued function $\mathbf{r}$ is the general antiderivative of $\mathbf{r}$ and represents the collection of all antiderivatives of $\mathbf{r}$.

The same reasoning that allows us to differentiate a vector-valued function componentwise applies to integrating as well. Recall that the integral of a sum is the sum of the integrals and also that we can remove constant factors from integrals. So, given $\mathbf{r}(t) = x(t)\mathbf{i} + y(t)\mathbf{j} + z(t)\mathbf{k}$, it follows that we can integrate componentwise. Expressed more formally,

> If $\mathbf{r}(t) = x(t)\mathbf{i} + y(t)\mathbf{j} + z(t)\mathbf{k}$, then
> $$\int \mathbf{r}(t)\,dt = \left(\int x(t)\,dt\right)\mathbf{i} + \left(\int y(t)\,dt\right)\mathbf{j} + \left(\int z(t)\,dt\right)\mathbf{k}.$$

In light of being able to integrate and differentiate componentwise with vector-valued functions, we can solve many problems that are analogous to those we encountered in single-variable calculus. For instance, recall problems where we were given an object moving along an axis with velocity function v and an initial position $s(0)$. In that context, we were able to differentiate v in order to find acceleration, and integrate v and use the initial condition in order to find the position function s. In the following activity, we explore similar ideas with vector-valued functions.

Activity 9.30.

Suppose a moving object in space has its velocity given by

$$\mathbf{v}(t) = (-2\sin(2t))\mathbf{i} + (2\cos(t))\mathbf{j} + \left(1 - \frac{1}{1+t}\right)\mathbf{k}.$$

A graph of the position of the object for times t in $[-0.5, 3]$ is shown in Figure 9.58. Suppose further that the object is at the point $(1.5, -1, 0)$ at time $t = 0$.

(a) Determine $\mathbf{a}(t)$, the acceleration of the object at time t.

(b) Determine $\mathbf{r}(t)$, position of the object at time t.

(c) Compute and sketch the position, velocity, and acceleration vectors of the object at time $t = 1$, using Figure 9.58.

(d) Finally, determine the vector equation for the tangent line, $\mathbf{L}(t)$, that is tangent to the position curve at $t = 1$.

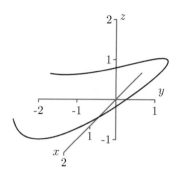

Figure 9.58: The position graph for the function in Activity 9.30.

◁

Projectile Motion

Any time that an object is launched into the air with a given velocity and launch angle, the path the object travels is determined almost exclusively by the force of gravity. Whether in sports such as archery or shotput, in military applications with artillery, or in important fields like firefighting, it is important to be able to know when and where a launched projectile will land. We can use our knowledge of vector-valued functions in order to completely determine the path traveled by an object that is launched from a given position at a given angle from the horizontal with a given initial velocity.

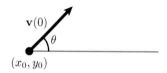

Figure 9.59: Projectile motion.

Assume we fire a projectile from a launcher and the only force acting on the fired object is the force of gravity pulling down on the object. That is, we assume no effect due to spin, wind, or air resistance. With these assumptions, the motion of the object will be planar, so we can also assume that the motion occurs in two-dimensional space. Suppose we launch the object from an initial position (x_0, y_0) at an angle θ with the positive x-axis as illustrated in Figure 9.59, and that we fire the object with an initial speed of $v_0 = |\mathbf{v}(0)|$, where $\mathbf{v}(t)$ is the velocity vector of the object at time t. Assume g is the positive constant acceleration force due to gravity, which acts to pull the fired object toward the ground (in the negative y direction). Note particularly that there is no external force acting on the object to move it in the x direction.

We first observe that since gravity only acts in the downward direction and that the acceleration due to gravity is constant, the acceleration vector is $\langle 0, -g \rangle$. That is, $\mathbf{a}(t) = \langle 0, -g \rangle$. We may use this fact about acceleration, together with the initial position and initial velocity in order to fully determine the position $\mathbf{r}(t)$ of the object at time t. In Exercise 5, you can work through the details to show that the following general formula holds.

If an object is launched from a point (x_0, y_0) with initial velocity v_0 at an angle θ with the horizontal, then the position of the object at time t is given by

$$\mathbf{r}(t) = \left\langle v_0 \cos(\theta)t + x_0, -\frac{g}{2}t^2 + v_0 \sin(\theta)t + y_0 \right\rangle.$$

This assumes that the only force acting on the object is the acceleration g due to gravity.

Summary

- If $\mathbf{r}$ is a vector-valued function, then the derivative of $\mathbf{r}$ is defined by

$$\mathbf{r}'(t) = \lim_{h \to 0} \frac{\mathbf{r}(t+h) - \mathbf{r}(t)}{h}$$

for those values of t at which the limit exists, and is computed componentwise by the formula

$$\mathbf{r}'(t) = x'(t)\mathbf{i} + y'(t)\mathbf{j} + z'(t)\mathbf{k}$$

for those values of t at which x, y, and z are differentiable, where $\mathbf{r}(t) = x(t)\mathbf{i} + y(t)\mathbf{j} + z(t)\mathbf{k}$.

- The derivative $\mathbf{r}'(t)$ of the vector-valued function $\mathbf{r}$ tells us the instantaneous rate of change of $\mathbf{r}$ with respect to time, t, which can be interpreted as a direction vector for the line tangent to the graph of $\mathbf{r}$ at the point $\mathbf{r}(t)$, or also as the instantaneous velocity of an object traveling along the graph defined by $\mathbf{r}(t)$ at time t.

- An antiderivative of $\mathbf{r}$ is a vector-valued function $\mathbf{R}$ such that $\mathbf{R}'(t) = \mathbf{r}(t)$. The indefinite integral $\int \mathbf{r}(t)\, dt$ of a vector-valued function $\mathbf{r}$ is the general antiderivative of $\mathbf{r}$ (which is a collection of all of the antiderivatives of $\mathbf{r}$, with any two antiderivatives differing by at most a constant vector). Moreover, if $\mathbf{r}(t) = x(t)\mathbf{i} + y(t)\mathbf{j} + z(t)\mathbf{k}$, then

$$\int \mathbf{r}(t)\, dt = \left(\int x(t)\, dt \right) \mathbf{i} + \left(\int y(t)\, dt \right) \mathbf{j} + \left(\int z(t)\, dt \right) \mathbf{k}.$$

- If an object is launched from a point (x_0, y_0) with initial velocity v_0 at an angle θ with the horizontal, then the position of the object at time t is given by

$$\mathbf{r}(t) = \left\langle v_0 \cos(\theta)t + x_0, -\frac{g}{2}t^2 + v_0 \sin(\theta)t + y_0 \right\rangle,$$

provided that that the only force acting on the object is the acceleration g due to gravity.

Exercises

1. Compute the derivative of each of the following functions in two different ways: (1) use the rules provided in the theorem stated just after Activity 9.27, and (2) rewrite each given function so that it is stated as a single function (either a scalar function or a vector-valued function with three components), and differentiate component-wise. Compare your answers to ensure that they are the same.

    (a) $\mathbf{r}(t) = \sin(t)\langle 2t, t^2, \arctan(t)\rangle$

    (b) $\mathbf{s}(t) = \mathbf{r}(2^t)$, where $\mathbf{r}(t) = \langle t+2, \ln(t), 1\rangle$.

    (c) $\mathbf{r}(t) = \langle \cos(t), \sin(t), t\rangle \cdot \langle -\sin(t), \cos(t), 1\rangle$

    (d) $\mathbf{r}(t) = \langle \cos(t), \sin(t), t\rangle \times \langle -\sin(t), \cos(t), 1\rangle$

2. Consider the two vector-valued functions given by

$$\mathbf{r}(t) = \left\langle t+1, \cos\left(\frac{\pi}{2}t\right), \frac{1}{1+t} \right\rangle$$

and

$$\mathbf{w}(s) = \left\langle s^2, \sin\left(\frac{\pi}{2}s\right), s \right\rangle.$$

   (a) Determine the point of intersection of the curves generated by $\mathbf{r}(t)$ and $\mathbf{w}(s)$. To do so, you will have to find values of a and b that result in $\mathbf{r}(a)$ and $\mathbf{w}(b)$ being the same vector.

   (b) Use the value of a you determined in (a) to find a vector form of the tangent line to $\mathbf{r}(t)$ at the point where $t = a$.

   (c) Use the value of b you determined in (a) to find a vector form of the tangent line to $\mathbf{w}(s)$ at the point where $s = b$.

   (d) Suppose that $z = f(x, y)$ is a function that generates a surface in three-dimensional space, and that the curves generated by $\mathbf{r}(t)$ and $\mathbf{w}(s)$ both lie on this surface. Note particularly that the point of intersection you found in (a) lies on this surface. In addition, observe that the two tangent lines found in (b) and (c) both lie in the tangent plane to the surface at the point of intersection. Use your preceding work to determine the equation of this tangent plane.

3. In this exercise, we determine the equation of a plane tangent to the surface defined by $f(x, y) = \sqrt{x^2 + y^2}$ at the point $(3, 4, 5)$.

   (a) Find a parameterization for the $x = 3$ trace of f. What is a direction vector for the line tangent to this trace at the point $(3, 4, 5)$?

   (b) Find a parameterization for the $y = 4$ trace of f. What is a direction vector for the line tangent to this trace at the point $(3, 4, 5)$?

   (c) The direction vectors in parts (a) and (b) form a plane containing the point $(3, 4, 5)$. What is a normal vector for this plane?

(d) Use your work in parts (a), (b), and (c) to deterring an equation for the tangent plane. Then, use appropriate technology to draw the graph of f and the plane you determined on the same set of axes. What do you observe? (We will discuss tangent planes in more detail in Chapter 10.)

4. For each given function **r**, determine $\int \mathbf{r}(t)\,dt$. In addition, recalling the Fundamental Theorem of Calculus for functions of a single variable, also evaluate $\int_0^1 \mathbf{r}(t)\,dt$ for each given function **r**. Is the resulting quantity a scalar or a vector? What does it measure?

   (a) $\mathbf{r}(t) = \left\langle \cos(t), \dfrac{1}{t+1}, te^t \right\rangle$

   (b) $\mathbf{r}(t) = \langle \cos(3t), \sin(2t), t \rangle$

   (c) $\mathbf{r}(t) = \left\langle \dfrac{t}{1+t^2}, te^{t^2}, \dfrac{1}{1+t^2} \right\rangle$

5. In this exercise, we develop the formula for the position function of a projectile that has been launched at an initial speed of $|\mathbf{v}_0|$ and a launch angle of θ. Recall that $\mathbf{a}(t) = \langle 0, -g \rangle$ is the constant acceleration of the projectile at any time t.

   (a) Find all velocity vectors for the given acceleration vector **a**. When you anti-differentiate, remember that there is an arbitrary constant that arises in each component.

   (b) Use the given information about initial speed and launch angle to find $\mathbf{v}_0$, the initial velocity of the projectile. You will want to write the vector in terms of its components, which will involve $\sin(\theta)$ and $\cos(\theta)$.

   (c) Next, find the specific velocity vector function **v** for the projectile. That is, combine your work in (a) and (b) in order to determine expressions in terms of $|\mathbf{v}_0|$ and θ for the constants that arose when integrating.

   (d) Find all possible position vectors for the velocity vector $\mathbf{v}(t)$ you determined in (c).

   (e) Let $\mathbf{r}(t)$ denote the position vector function for the given projectile. Use the fact that the object is fired from the position (x_0, y_0) to show it follows that
   $$\mathbf{r}(t) = \left\langle |\mathbf{v}_0|\cos(\theta)t + x_0,\ -\dfrac{g}{2}t^2 + |\mathbf{v}_0|\sin(\theta)t + y_0 \right\rangle.$$

6. A *central force* is one that acts on an object so that the force **F** is parallel to the object's position **r**. Since Newton's Second Law says that an object's acceleration is proportional to the force exerted on it, the acceleration **a** of an object moving under a central force will be parallel to its position **r**. For instance, the Earth's acceleration due to the gravitational force that the sun exerts on the Earth is parallel to the Earth's position vector as shown in Figure 9.60.

   (a) If an object of mass m is moving under a central force, the angular momentum vector is defined to be $\mathbf{L} = m\mathbf{r} \times \mathbf{v}$. Assuming the mass is constant, show that the angular momentum is constant by showing that
   $$\dfrac{d\mathbf{L}}{dt} = \mathbf{0}.$$

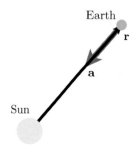

Figure 9.60: A central force.

(b) Explain why $\mathbf{L} \cdot \mathbf{r} = 0$.

(c) Explain why we may conclude that the object is constrained to lie in the plane passing through the origin and perpendicular to $\mathbf{L}$.

9.8 Arc Length and Curvature

Motivating Questions

In this section, we strive to understand the ideas generated by the following important questions:

- How can a definite integral be used to measure the length of a curve in 2- or 3-space?
- Why is arc length useful as a parameter?
- What is the curvature of a curve?

Introduction

Given a space curve, there are two natural geometric questions one might ask: how long is the curve and how much does it bend? In this section, we answer both questions by developing techniques for measuring the length of a space curve as well as its curvature.

Preview Activity 9.8. In earlier investigations, we have used integration to calculate quantities such as area, volume, mass, and work. We are now interested in determining the length of a space curve.

Consider the smooth curve in 3-space defined by the vector-valued function $\mathbf{r}$, where

$$\mathbf{r}(t) = \langle x(t), y(t), z(t) \rangle = \langle \cos(t), \sin(t), t \rangle$$

for t in the interval $[0, 2\pi]$. Pictures of the graph of $\mathbf{r}$ are shown in Figure 9.61. We will use the integration process to calculate the length of this curve. In this situation we partition the interval $[0, 2\pi]$ into n subintervals of equal length and let $0 = t_0 < t_1 < t_2 < \cdots < t_n = b$ be the endpoints of the subintervals. We then approximate the length of the curve on each subinterval with some related quantity that we can compute. In this case, we approximate the length of the curve on each subinterval with the length of the segment connecting the endpoints. Figure 9.61 illustrates the process in three different instances using increasing values of n.

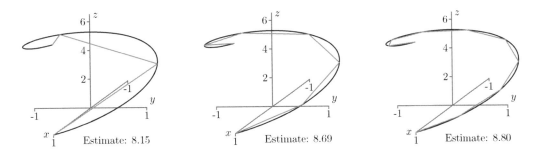

Figure 9.61: Approximating the length of the curve with $n = 3$, $n = 6$, and $n = 9$.

(a) Write a formula for the length of the line segment that connects the endpoints of the curve on the ith subinterval $[t_{i-1}, t_i]$. (This length is our approximation of the length of the curve on this interval.)

(b) Use your formula in part (a) to write a sum that adds all of the approximations to the lengths on each subinterval.

(c) What do we need to do with the sum in part (b) in order to obtain the exact value of the length of the graph of $\mathbf{r}(t)$ on the interval $[0, 2\pi]$?

⋈

Arc Length

Consider a smooth curve in 3-space that is parametrically described by the vector-valued function $\mathbf{r}$ defined by $\mathbf{r}(t) = \langle x(t), y(t), z(t) \rangle$. Preview Activity 9.8 shows that to approximate the length of the curve defined by $\mathbf{r}(t)$ as the values of t run over an interval $[a, b]$, we partition the interval $[a, b]$ into n subintervals of equal length Δt, with $a = t_0 < t_1 < \cdots < t_n = b$ as the endpoints of the subintervals. On each subinterval, we approximate the length of the curve by the length of the line segment connecting the endpoints. The points on the curve corresponding to $t = t_{i-1}$ and $t = t_i$ are $(x(t_{i-1}), y(t_{i-1}), z(t_{i-1}))$ and $(x(t_i), y(t_i), z(t_i))$, respectively, so the length of the line segment connecting these points is

$$\sqrt{(x(t_i) - x(t_{i-1}))^2 + (y(t_i) - y(t_{i-1}))^2 + (z(t_i) - z(t_{i-1}))^2}.$$

Now we add all of these approximations together to obtain an approximation to the length L of the curve:

$$L \approx \sum_{i=1}^{n} \sqrt{(x(t_i) - x(t_{i-1}))^2 + (y(t_i) - y(t_{i-1}))^2 + (z(t_i) - z(t_{i-1}))^2}.$$

We now want to take the limit of this sum as n goes to infinity, but in its present form it might be difficult to see how. We first introduce Δt by multiplying by $\frac{\Delta t}{\Delta t}$, and see that

$$L \approx \sum_{i=1}^{n} \sqrt{(x(t_i) - x(t_{i-1}))^2 + (y(t_i) - y(t_{i-1}))^2 + (z(t_i) - z(t_{i-1}))^2}$$

$$= \sum_{i=1}^{n} \sqrt{(x(t_i) - x(t_{i-1}))^2 + (y(t_i) - y(t_{i-1}))^2 + (z(t_i) - z(t_{i-1}))^2} \frac{\Delta t}{\Delta t}$$

$$= \sum_{i=1}^{n} \sqrt{(x(t_i) - x(t_{i-1}))^2 + (y(t_i) - y(t_{i-1}))^2 + (z(t_i) - z(t_{i-1}))^2} \frac{\Delta t}{\sqrt{(\Delta t)^2}}$$

9.8. ARC LENGTH AND CURVATURE

To get the difference quotients under the radical, we use properties of the square root function to see further that

$$L \approx \sum_{i=1}^{n} \sqrt{[(x(t_i) - x(t_{i-1}))^2 + (y(t_i) - y(t_{i-1}))^2 + (z(t_i) - z(t_{i-1})^2] \frac{1}{(\Delta t)^2}} \Delta t$$

$$= \sum_{i=1}^{n} \sqrt{\left(\frac{x(t_i) - x(t_{i-1})}{\Delta t}\right)^2 + \left(\frac{y(t_i) - y(t_{i-1})}{\Delta t}\right)^2 + \left(\frac{z(t_i) - z(t_{i-1})}{\Delta t}\right)^2} \Delta t.$$

Recall that as $n \to \infty$ we also have $\Delta t \to 0$. Since

$$\lim_{\Delta t \to 0} \frac{x(t_i) - x(t_{i-1})}{\Delta t} = x'(t), \quad \lim_{\Delta t \to 0} \frac{y(t_i) - y(t_{i-1})}{\Delta t} = y'(t), \quad \text{and} \quad \lim_{\Delta t \to 0} \frac{z(t_i) - z(t_{i-1})}{\Delta t} = z'(t)$$

we see that

$$L = \lim_{n \to \infty} \sum_{i=1}^{n} \sqrt{\left(\frac{x(t_i) - x(t_{i-1})}{\Delta t}\right)^2 + \left(\frac{y(t_i) - y(t_{i-1})}{\Delta t}\right)^2 + \left(\frac{z(t_i) - z(t_{i-1})}{\Delta t}\right)^2} \Delta t$$

$$= \lim_{\Delta t \to 0} \sum_{i=1}^{n} \sqrt{\left(\frac{x(t_i) - x(t_{i-1})}{\Delta t}\right)^2 + \left(\frac{y(t_i) - y(t_{i-1})}{\Delta t}\right)^2 + \left(\frac{z(t_i) - z(t_{i-1})}{\Delta t}\right)^2} \Delta t$$

$$= \int_a^b \sqrt{(x'(t))^2 + (y'(t))^2 + (z'(t))^2} \, dt. \tag{9.13}$$

Noting further that

$$|\mathbf{r}'(t)| = \sqrt{(x'(t))^2 + (y'(t))^2 + (z'(t))^2},$$

we can rewrite (9.13) in a more succinct form as follows.

If $\mathbf{r}(t)$ defines a smooth curve C on an interval $[a, b]$, then the **length** L of C is given by

$$L = \int_a^b |\mathbf{r}'(t)| \, dt. \tag{9.14}$$

[handwritten: vector valued — in order to find arc length you need to take the integral of a vector-valued func.]

Note that formula (9.14) applies to curves in any dimensional space. Moreover, this formula has a natural interpretation: if $\mathbf{r}(t)$ records the position of a moving object, then $\mathbf{r}'(t)$ is the object's velocity and $|\mathbf{r}'(t)|$ its speed. Formula (9.14) says that we simply integrate the speed of an object traveling over the curve to find the distance traveled by the object, which is the same as the length of the curve, just as in one-variable calculus.

Activity 9.31.

Here we calculate the arc length of two familiar curves.

(a) Use Equation (9.14) to calculate the circumference of a circle of radius r.

(b) Find the exact length of the spiral defined by $\mathbf{r}(t) = \langle \cos(t), \sin(t), t \rangle$ on the interval $[0, 2\pi]$.

◁

We can adapt the arc length formula to curves in 2-space that define y as a function of x as the following activity shows.

Activity 9.32.

Let $y = f(x)$ define a smooth curve in 2-space. Parameterize this curve and use Equation (9.14) to show that the length of the curve defined by f on an interval $[a, b]$ is

$$\int_a^b \sqrt{1 + [f'(t)]^2}\, dt.$$

◁

Parameterizing With Respect To Arc Length

In addition to helping us to find the length of space curves, the expression for the length of a curve enables us to find a natural parametrization of space curves in terms of arc length, as we now explain.

Shown below in Figure 9.62 is a portion of the parabola $y = x^2/2$. Of course, this space curve may be parametrized by the vector-valued function $\mathbf{r}$ defined by $\mathbf{r}(t) = \langle t, t^2/2 \rangle$ as shown on the left, where we see the location at a few different times t. Notice that the points are not equally spaced on the curve.

A more natural parameter describing the points along the space curve is the distance traveled s as we move along the parabola starting at the origin. For instance, the right side of Figure 9.62 shows the points corresponding to various values of s. We call this an *arc length parametrization*.

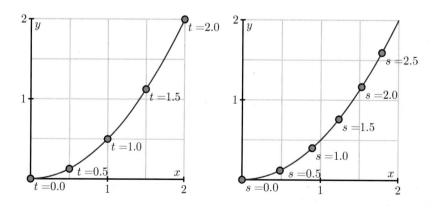

Figure 9.62: The parametrization $\mathbf{r}(t)$ (left) and a reparametrization by arc length.

9.8. ARC LENGTH AND CURVATURE

To see that this is a more natural parametrization, consider an interstate highway cutting across a state. One way to parametrize the curve defined by the highway is to drive along the highway and record our position at every time, thus creating a function **r**. If we encounter an accident or road construction, however, this parametrization might not be at all relevant to another person driving the same highway. An arc length parametrization, however, is like using the mile markers on the side of road to specify our position on the highway. If we know how far we've traveled along the highway, we know exactly where we are.

If we begin with a parametrization of a space curve, we can modify it to find an arc length parametrization, as we now describe. Suppose that the curve is parametrized by the vector-valued function $\mathbf{r} = \mathbf{r}(t)$ where t is in the interval $[a, b]$. We define the parameter s through the function

$$s = L(t) = \int_a^t \sqrt{(x'(w))^2 + (y'(w))^2 + (z'(w))^2}\, dw,$$

which measures the length along the curve from $\mathbf{r}(a)$ to $\mathbf{r}(t)$.

The Fundamental Theorem of Calculus shows us that

$$\frac{ds}{dt} = L'(t) = \sqrt{(x'(t))^2 + (y'(t))^2 + (z'(t))^2} = |\mathbf{r}'(t)| \tag{9.15}$$

and so

$$L(t) = \int_a^t \left|\frac{d}{dt}\mathbf{r}(w)\right| dw.$$

If we assume that $\mathbf{r}'(t)$ is never 0, then $L'(t) > 0$ for all t and $s = L(t)$ is always increasing. This should seem reasonable: unless we stop, the distance traveled along the curve increases as we move along the curve.

Since $s = L(t)$ is an increasing function, it is invertible, which means we may view the time t as a function of the distance traveled; that is, we have the relationship $t = L^{-1}(s)$. We then obtain the arc length parametrization by composing $\mathbf{r}(t)$ with $t = L^{-1}(s)$ to obtain $\mathbf{r}(s)$. Let's illustrate this with an example.

Example 9.1. Consider a circle of radius 5 in 2-space centered at the origin. We know that we can parameterize this circle as

$$\mathbf{r}(t) = \langle 5\cos(t), 5\sin(t)\rangle,$$

where t runs from 0 to 2π. We see that $\mathbf{r}'(t) = \langle -5\sin(t), 5\cos(t)\rangle$, and hence $|\mathbf{r}'(t)| = 5$. It then follows that

$$s = L(t) = \int_0^t |\mathbf{r}'(w)|\, dw = \int_0^t 5\, dw = 5t.$$

Since $s = L(t) = 5t$, we may solve for t in terms of s to obtain $t = L^{-1}(s) = s/5$. We then find the arc length parametrization by composing

$$\mathbf{r}(s) = \mathbf{r}(L^{-1}(s)) = \left\langle 5\cos\left(\frac{s}{5}\right), 5\sin\left(\frac{s}{5}\right)\right\rangle.$$

More generally, for a circle of radius a centered at the origin, a similar computation shows that

$$\mathbf{r}(s) = \left\langle a\cos\left(\frac{s}{a}\right), a\sin\left(\frac{s}{a}\right) \right\rangle \qquad (9.16)$$

is an arc length parametrization.

Notice that equation (9.15) shows that

$$\frac{d\mathbf{r}}{dt} = \frac{d\mathbf{r}}{ds}\frac{ds}{dt} = \frac{d\mathbf{r}}{ds}|\mathbf{r}'(t)|,$$

so

$$\left|\frac{d\mathbf{r}}{ds}\right| = \left|\frac{1}{|\mathbf{r}'(t)|}\frac{d\mathbf{r}}{dt}\right| = 1,$$

which means that we move along the curve with unit speed when we parameterize by arc length. This is clearly seen in Example 9.1 where $|\mathbf{r}'(s)| = 1$. It follows that the parameter s is the distance traveled along the curve, as shown by:

$$L(s) = \int_0^s \left|\frac{d}{ds}\mathbf{r}(w)\right| dw = \int_0^s 1\, dw = s.$$

Activity 9.33.

In this activity we parameterize a line in 2-space in terms of arc length. Consider the line with parametric equations

$$x(t) = x_0 + at \quad \text{and} \quad y(t) = y_0 + bt.$$

(a) To write t in terms of s, evaluate the integral

$$s = L(t) = \int_0^t \sqrt{(x'(w))^2 + (y'(w))^2}\, dw$$

to determine the length of the line from time 0 to time t.

(b) Use the formula from (a) for s in terms of t to write t in terms of s. Then explain why a parameterization of the line in terms of arc length is

$$x(s) = x_0 + \frac{a}{\sqrt{a^2+b^2}}s \quad \text{and} \quad y(s) = y_0 + \frac{b}{\sqrt{a^2+b^2}}s. \qquad (9.17)$$

◁

A little more complicated example is the following.

Example 9.2. Let us parameterize the curve defined by

$$\mathbf{r}(t) = \left\langle t^2, \frac{8}{3}t^{3/2}, 4t \right\rangle$$

9.8. ARC LENGTH AND CURVATURE

for $t \geq 0$ in terms of arc length. To write t in terms of s we find s in terms of t:

$$\begin{aligned}
s(t) &= \int_0^t \sqrt{(x'(w))^2 + (y'(w))^2 + (z'(w))^2}\, dw \\
&= \int_0^t \sqrt{(2w)^2 + (4w^{1/2})^2 + (4)^2}\, dw \\
&= \int_0^t \sqrt{4w^2 + 16w + 16}\, dw \\
&= 2\int_0^t \sqrt{(w+2)^2}\, dw \\
&= 2\int_0^t w + 2\, dw \\
&= (w^2 + 4w)\Big|_0^t \\
&= t^2 + 4t.
\end{aligned}$$

Since $t \geq 0$, we can solve the equation $s = t^2 + 4t$ (or $t^2 + 4t - s = 0$) for t to obtain $t = \frac{-4 + \sqrt{16 + 4s}}{2} = -2 + \sqrt{4+s}$. So we can parameterize our curve in terms of arc length by

$$\mathbf{r}(s) = \left\langle \left(-2 + \sqrt{4+s}\right)^2, \frac{8}{3}\left(-2 + \sqrt{4+s}\right)^{3/2}, 4\left(-2 + \sqrt{4+s}\right) \right\rangle.$$

These examples illustrate a general method. Of course, evaluating an arc length integral and finding a formula for the inverse of a function can be difficult, so while this process is theoretically possible, it is not always practical to parameterize a curve in terms of arc length. However, we can guarantee that such a parameterization exists, and this observation plays an important role in the next section.

Curvature

For a smooth space curve, the *curvature* measures how fast the curve is bending or changing direction at a given point. For example, we expect that a line should have zero curvature everywhere, while a circle (which is bending the same at every point) should have constant curvature. Circles with larger radii should have smaller curvatures.

To measure the curvature, we first need to describe the direction of the curve at a point. We may do this using a continuously varying tangent vector to the curve, as shown in 9.63. The direction of the curve is then determined by the angle ϕ this tangent vector makes with a horizontal vector, as shown in 9.64.

Informally speaking, the curvature will be the rate at which the angle ϕ is changing as we move along the curve. Of course, this rate of change will depend on how we move along the curve; if we move with a greater speed along the curve, then ϕ will change more rapidly. This is why the

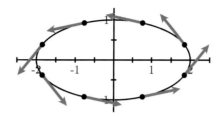
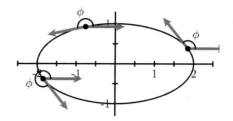

Figure 9.63: Tangent vectors to an ellipse.   Figure 9.64: Angles of tangent vectors.

speed limit is sometimes lowered when we enter a curve on a highway. In other words, the rate of change of ϕ will depend on the parametrization we use to describe the space curve. To eliminate this dependence on the parametrization, we choose to work with an arc length parametrization $\mathbf{r}(s)$, which means we move along the curve with unit speed.

Using an arc length parametrization $\mathbf{r}(s)$, we define the tangent vector $\mathbf{T}(s) = \mathbf{r}'(s)$, and note that $|\mathbf{T}(s)| = 1$; that is, $\mathbf{T}(s)$ is a unit tangent vector. We then have $\mathbf{T}(s) = \langle \cos(\phi(s)), \sin(\phi(s)) \rangle$, which means that

$$\frac{d\mathbf{T}}{ds} = \left\langle -\sin(\phi(s))\frac{d\phi}{ds},\ \cos(\phi(s))\frac{d\phi}{ds} \right\rangle = \langle -\sin(\phi(s)),\ \cos(\phi(s))\rangle \frac{d\phi}{ds}.$$

Therefore

$$\left|\frac{d\mathbf{T}}{ds}\right| = |\langle -\sin(\phi(s)),\ \cos(\phi(s))\rangle| \left|\frac{d\phi}{ds}\right| = \left|\frac{d\phi}{ds}\right|$$

This observation leads us to adopt

Definition 9.12. If C is a smooth space curve and s is an arc length parameter for C, then the **curvature**, κ,[14] of C is

$$\kappa = \kappa(s) = \left|\frac{d\mathbf{T}}{ds}\right|.$$

Example 9.3. We should expect that the curvature of a line is 0 everywhere. To show that our definition of curvature measures this correctly in 2-space, recall that (9.17) gives us the arc length parameterization

$$x(s) = x_0 + \frac{a}{\sqrt{a^2 + b^2}} s \quad \text{and} \quad y(s) = y_0 + \frac{b}{\sqrt{a^2 + b^2}} s$$

of a line. So the unit tangent vector is

$$\mathbf{T}(s) = \left\langle \frac{a}{\sqrt{a^2 + b^2}},\ \frac{b}{\sqrt{a^2 + b^2}} \right\rangle$$

9.8. ARC LENGTH AND CURVATURE

and, since $\mathbf{T}(s)$ is constant, we have

$$\kappa = \left|\frac{d\mathbf{T}}{ds}\right| = 0$$

as expected.

Activity 9.34.

Recall that an arc length parameterization of a circle in 2-space of radius a centered at the origin is, from (9.16),

$$\mathbf{r}(s) = \left\langle a\cos\left(\frac{s}{a}\right), a\sin\left(\frac{s}{a}\right)\right\rangle.$$

Show that the curvature of this circle is the constant $\frac{1}{a}$. What can you say about the relationship between the size of the radius of a circle and the value of its curvature? Why does this make sense?

◁

The definition of curvature relies on our ability to parameterize curves in terms of arc length. Since we have seen that finding an arc length parametrization can be difficult, we would like to be able to express the curvature in terms of a more general parametrization $\mathbf{r}(t)$.

To begin, we need to describe the vector $\mathbf{T}$, which is a vector tangent to the curve having unit length. Of course, the velocity vector $\mathbf{r}'(t)$ is tangent to the curve; we simply need to normalize its length to be one. This means that we may take

$$\mathbf{T}(t) = \frac{\mathbf{r}'(t)}{|\mathbf{r}'(t)|}.$$

Then the curvature of the curve defined by $\mathbf{r}$ is

$$\kappa = \left|\frac{d\mathbf{T}}{ds}\right|$$
$$= \left|\frac{d\mathbf{T}}{dt}\frac{dt}{ds}\right|$$
$$= \frac{\left|\frac{d\mathbf{T}}{dt}\right|}{\left|\frac{ds}{dt}\right|}$$
$$= \frac{|\mathbf{T}'(t)|}{|\mathbf{r}'(t)|}.$$

This last formula allows us to use any parameterization of a curve to calculate its curvature.

There is another useful formula, given below, whose derivation is left for the exercises.

> If **r** is a vector-valued function defining a smooth space curve C, and if $\mathbf{r}'(t)$ is not zero and if $\mathbf{r}''(t)$ exists, then the curvature κ of C satisfies
>
> - $\kappa = \kappa(t) = \frac{|\mathbf{T}'(t)|}{|\mathbf{r}'(t)|}$
>
> - $\kappa = \frac{|\mathbf{r}'(t) \times \mathbf{r}''(t)|}{|\mathbf{r}'(t)|^3}$.

Activity 9.35.

Use one of the two formulas for κ in terms of t to help you answer the following questions.

(a) The ellipse $\frac{x^2}{a^2} + \frac{y^2}{b^2} = 1$ has parameterization

$$\mathbf{r}(t) = \langle a\cos(t), b\sin(t) \rangle.$$

Find the curvature of the ellipse. Assuming $0 < b < a$, at what points is the curvature the greatest and at what points is the curvature the smallest? Does this agree with your intuition?

(b) The standard helix has parameterization $\mathbf{r}(t) = \cos(t)\mathbf{i} + \sin(t)\mathbf{j} + t\mathbf{k}$. Find the curvature of the helix. Does the result agree with your intuition?

◁

The curvature has another interpretation. Recall that the tangent line to a curve at a point is the line that best approximates the curve at that point. The curvature at a point on a curve describes the *circle* that best approximates the curve at that point. Remembering that a circle of radius a has curvature $1/a$, then the circle that best approximates the curve near a point on a curve whose curvature is κ has radius $1/\kappa$ and will be tangent to the tangent line at that point. This circle, called the *osculating circle* of the curve at the point, is shown in Figure 9.65 for a portion of a parabola.

Summary

- The integration process shows that the length L of a smooth curve defined by $\mathbf{r}(t)$ on an interval $[a, b]$ is

$$L = \int_a^b |\mathbf{r}'(t)|\, dt.$$

- Arc length is useful as a parameter because when we parameterize with respect to arc length, we eliminate the role of speed in our calculation of curvature and the result is a measure that depends only on the geometry of the curve and not on the parameterization of the curve.

- We define the curvature κ of a curve in 2- or 3-space to be the rate of change of the unit tangent vector with respect to arc length, or

$$\kappa = \left|\frac{d\mathbf{T}}{ds}\right|.$$

9.8. ARC LENGTH AND CURVATURE

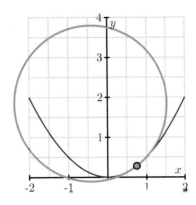

Figure 9.65: The osculating circle

Exercises

1. Consider the moving particle whose position at time t in seconds is given by the vector-valued function $\mathbf{r}$ defined by $\mathbf{r}(t) = 5t\mathbf{i} + 4\sin(3t)\mathbf{j} + 4\cos(3t)\mathbf{k}$. Use this function to answer each of the following questions.

   (a) Find the unit tangent vector, $\mathbf{T}(t)$, to the spacecurve traced by $\mathbf{r}(t)$ at time t. Write one sentence that explains what $\mathbf{T}(t)$ tells us about the particle's motion.

   (b) Determine the speed of the particle moving along the spacecurve with the given parameterization.

   (c) Find the exact distance traveled by the particle on the time interval $[0, \pi/3]$.

   (d) Find the average velocity of the particle on the time interval $[0, \pi/3]$.

   (e) Determine the parameterization of the given curve with respect to arc length.

2. Let $y = f(x)$ define a curve in the plane. We can consider this curve as a curve in three-space with z-coordinate 0.

   (a) Find a parameterization of the form $\mathbf{r}(t) = \langle x(t), y(t), z(t) \rangle$ of the curve $y = f(x)$ in three-space.

   (b) Use the formula
   $$\kappa = \frac{|\mathbf{r}'(t) \times \mathbf{r}''(t)|}{|\mathbf{r}'(t)|^3}$$
   to show that
   $$\kappa = \frac{|f''(x)|}{[1 + (f'(x))^2]^{3/2}}.$$

3. Consider the single variable function defined by $y = 4x^2 - x^3$.

(a) Find a parameterization of the form $\mathbf{r}(t) = \langle x(t), y(t)\rangle$ that traces the curve $y = 4x^2 - x^3$ on the interval from $x = -3$ to $x = 3$.

(b) Write a definite integral which, if evaluated, gives the exact length of the given curve from $x = -3$ to $x = 3$. Why is the integral difficult to evaluate exactly?

(c) Determine the curvature, $\kappa(t)$, of the parameterized curve. (Exercise 2 might be useful here.)

(d) Use appropriate technology to approximate the absolute maximum and minimum of $\kappa(t)$ on the parameter interval for your parameterization. Compare your results with the graph of $y = 4x^2 - x^3$. How do the absolute maximum and absolute minimum of $\kappa(t)$ align with the original curve?

4. Consider the standard helix parameterized by $\mathbf{r}(t) = \cos(t)\mathbf{i} + \sin(t)\mathbf{j} + t\mathbf{k}$.

   (a) Recall that the unit tangent vector, $\mathbf{T}(t)$, is the vector tangent to the curve at time t that points in the direction of motion and has length 1. Find $\mathbf{T}(t)$.

   (b) Explain why the fact that $|\mathbf{T}(t)| = 1$ implies that $\mathbf{T}$ and $\mathbf{T}'$ are orthogonal vectors for every value of t. (Hint: note that $\mathbf{T} \cdot \mathbf{T} = |\mathbf{T}|^2 = 1$, and compute $\frac{d}{dt}[\mathbf{T} \cdot \mathbf{T}]$.)

   (c) For the given function $\mathbf{r}$ with unit tangent vector $\mathbf{T}(t)$ (from (a)), determine $\mathbf{N}(t) = \frac{1}{|\mathbf{T}'(t)|}\mathbf{T}'(t)$.

   (d) What geometric properties does $\mathbf{N}(t)$ have? That is, how long is this vector, and how is it situated in comparison to $\mathbf{T}(t)$?

   (e) Let $\mathbf{B}(t) = \mathbf{T}(t) \times \mathbf{N}(t)$, and compute $\mathbf{B}(t)$ in terms of your results in (a) and (c).

   (f) What geometric properties does $\mathbf{B}(t)$ have? That is, how long is this vector, and how is it situated in comparison to $\mathbf{T}(t)$ and $\mathbf{N}(t)$?

   (g) Sketch a plot of the given helix, and compute and sketch $\mathbf{T}(\pi/2)$, $\mathbf{N}(\pi/2)$, and $\mathbf{B}(\pi/2)$.

5. In this exercise we verify the curvature formula
$$\kappa = \frac{|\mathbf{r}'(t) \times \mathbf{r}''(t)|}{|\mathbf{r}'(t)|^3}.$$

   (a) Explain why
   $$|\mathbf{r}'(t)| = \frac{ds}{dt}.$$

   (b) Use the fact that $\mathbf{T}(t) = \frac{\mathbf{r}'(t)}{|\mathbf{r}'(t)|}$ and $|\mathbf{r}'(t)| = \frac{ds}{dt}$ to explain why
   $$\mathbf{r}'(t) = \frac{ds}{dt}\mathbf{T}(t).$$

(c) The Product Rule shows that
$$\mathbf{r}''(t) = \frac{d^2s}{dt^2}\mathbf{T}(t) + \frac{ds}{dt}\mathbf{T}'(t).$$

Explain why
$$\mathbf{r}'(t) \times \mathbf{r}''(t) = \left(\frac{ds}{dt}\right)^2 (\mathbf{T}(t) \times \mathbf{T}'(t)).$$

(d) In Exercise 4 we showed that $|\mathbf{T}(t)| = 1$ implies that $\mathbf{T}(t)$ is orthogonal to $\mathbf{T}'(t)$ for every value of t. Explain what this tells us about $|\mathbf{T}(t) \times \mathbf{T}'(t)|$ and conclude that
$$|\mathbf{r}'(t) \times \mathbf{r}''(t)| = \left(\frac{ds}{dt}\right)^2 |\mathbf{T}'(t)|.$$

(e) Finally, use the fact that $\kappa = \frac{|\mathbf{T}'(t)|}{|\mathbf{r}'(t)|}$ to verify that
$$\kappa = \frac{|\mathbf{r}'(t) \times \mathbf{r}''(t)|}{|\mathbf{r}'(t)|^3}.$$

Chapter 10

Derivatives of Multivariable Functions

10.1 Limits

Motivating Questions

In this section, we strive to understand the ideas generated by the following important questions:

- What do we mean by the limit of a function f of two variables at a point (a, b)?

- What techniques can we use to show that a function of two variables does not have a limit at a point (a, b)?

- What does it mean for a function f of two variables to be continuous at a point (a, b)?

Introduction

In this section, we will study limits of functions of several variables, with a focus on limits of functions of two variables. In single variable calculus, we studied the notion of limit, which turned out to be a critical concept that formed the basis for the derivative and the definite integral. In this section we will begin to understand how the concept of limit for functions of two variables is similar to what we encountered for functions of a single variable. The limit will again be the fundamental idea in multivariable calculus, and we will use this notion of the limit of a function of several variables to define the important concept of differentiability later in this chapter. We have already seen its use in the derivatives of vector-valued functions in Section 9.7.

Let's begin by reviewing what we mean by the limit of a function of one variable. We say that a function f has a limit L as x approaches a provided that we can make the values $f(x)$ as close to L as we like by taking x sufficiently close (but not equal) to a. We denote this behavior by writing

$$\lim_{x \to a} f(x) = L.$$

Preview Activity 10.1. We investigate the limits of several different functions by working with tables and graphs.

(a) Consider the function f defined by
$$f(x) = 3 - x.$$
Complete the following table of values.

x	$f(x)$
-0.2	
-0.1	
0.0	
0.1	
0.2	

What does the table suggest regarding $\lim_{x \to 0} f(x)$?

(b) Explain how your results in (a) are reflected in Figure 10.1.

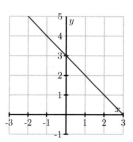

Figure 10.1: The graph of $f(x) = 3 - x$.

(c) Next, consider
$$g(x) = \frac{x}{|x|}.$$
Complete the following table of values near $x = 0$, the point at which g is not defined.

x	$g(x)$
-0.1	
-0.01	
-0.001	
0.001	
0.01	
0.1	

What does this suggest about $\lim_{x \to 0} g(x)$?

10.1. LIMITS

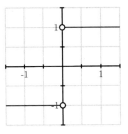

Figure 10.2: The graph of $g(x) = \frac{x}{|x|}$.

(d) Explain how your results in (c) are reflected in Figure 10.2.

(e) Now, let's examine a function of two variables. Let

$$f(x,y) = 3 - x - 2y$$

and complete the following table of values.

$x \backslash y$	-1	-0.1	0	0.1	1
-1					
-0.1					
0					
0.1					
1					

What does the table suggest about $\lim_{(x,y)\to(0,0)} f(x,y)$?

(f) Explain how your results in (e) are reflected in Figure 10.3. Compare this limit to the limit in part (a). How are the limits similar and how are they different?

(g) Finally, consider

$$g(x,y) = \frac{2xy}{x^2 + y^2},$$

which is not defined at $(0,0)$, and complete the following table of values of $g(x,y)$.

$x \backslash y$	-1	-0.1	0	0.1	1
-1					
-0.1					
0			—		
0.1					
1					

What does this suggest about $\lim_{(x,y)\to(0,0)} g(x,y)$?

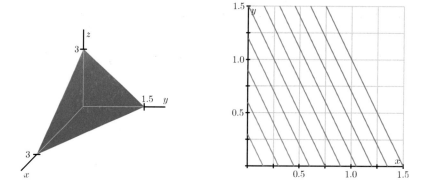

Figure 10.3: At left, the graph of $f(x, y) = 3 - x - 2y$; at right, its contour plot.

(h) Explain how your results are reflected in Figure 10.4. Compare this limit to the limit in part (b). How are the results similar and how are they different?

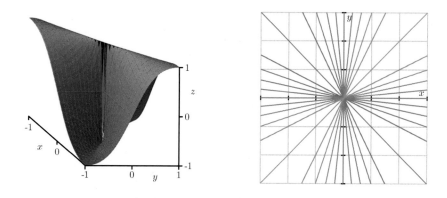

Figure 10.4: At left, the graph of $g(x, y) = \frac{2xy}{x^2+y^2}$; at right, its contour plot.

⋈

Limits of Functions of Two Variables

In Preview Activity 10.1, we recalled the notion of limit from single variable calculus and saw that a similar concept applies to functions of two variables. Though we will focus on functions of two variables, for the sake of discussion, all the ideas we establish here are valid for functions of any number of variables. In a natural followup to our work in Preview Activity 10.1, we now formally

10.1. LIMITS

define what it means for a function of two variables to have a limit at a point.

Definition 10.1. Given a function $f = f(x, y)$, we say that *f has limit L as (x, y) approaches (a, b)* provided that we can make $f(x, y)$ as close to L as we like by taking (x, y) sufficiently close (but not equal) to (a, b). We write

$$\lim_{(x,y) \to (a,b)} f(x, y) = L.$$

To investigate the limit of a single variable function, $\lim_{x \to a} f(x)$, we often consider the behavior of f as x approaches a from the right and from the left. Similarly, we may investigate limits of two-variable functions, $\lim_{(x,y) \to (a,b)} f(x, y)$ by considering the behavior of f as (x, y) approaches (a, b) from various directions. This situation is more complicated because there are infinitely many ways in which (x, y) may approach (a, b). In the next activity, we see how it is important to consider a variety of those paths in investigating whether or not a limit exists.

Activity 10.1.

Consider the function f, defined by

$$f(x, y) = \frac{y}{\sqrt{x^2 + y^2}},$$

whose graph is shown below in Figure 10.5

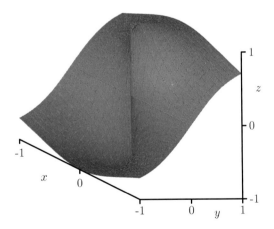

Figure 10.5: The graph of $f(x, y) = \frac{y}{\sqrt{x^2 + y^2}}$.

(a) Is f defined at the point $(0, 0)$? What, if anything, does this say about whether f has a limit at the point $(0, 0)$?

(b) Values of f (to three decimal places) at several points close to $(0, 0)$ are shown in the table below.

x\y	-1	-0.1	0	0.1	1
−1	−0.707	—	0	—	0.707
−0.1	—	−0.707	0	0.707	—
0	−1	−1	—	1	1
0.1	—	−0.707	0	0.707	—
1	−0.707	—	0	—	0.707

Based on these calculations, state whether f has a limit at $(0,0)$ and give an argument supporting your statement. (Hint: The blank spaces in the table are there to help you see the patterns.)

(c) Now let's consider what happens if we restrict our attention to the x-axis; that is, consider what happens when $y = 0$. What is the behavior of $f(x,0)$ as $x \to 0$? If we approach $(0,0)$ by moving along the x-axis, what value do we find as the limit?

(d) What is the behavior of f along the line $y = x$ when $x > 0$; that is, what is the value of $f(x,x)$ when $x > 0$? If we approach $(0,0)$ by moving along the line $y = x$ in the first quadrant (thus considering $f(x,x)$ as $x \to 0$, what value do we find as the limit?

(e) In general, if $\lim_{(x,y) \to (0,0)} f(x,y) = L$, then $f(x,y)$ approaches L as (x,y) approaches $(0,0)$, regardless of the path we take in letting $(x,y) \to (0,0)$. Explain what the last two parts of this activity imply about the existence of $\lim_{(x,y) \to (0,0)} f(x,y)$.

(f) Shown below in Figure 10.6 is a set of contour lines of the function f. What is the behavior of $f(x,y)$ as (x,y) approaches $(0,0)$ along any straight line? How does this observation reinforce your conclusion about the existence of $\lim_{(x,y) \to (0,0)} f(x,y)$ from the previous part of this activity? (Hint: Use the fact that a non-vertical line has equation $y = mx$ for some constant m.)

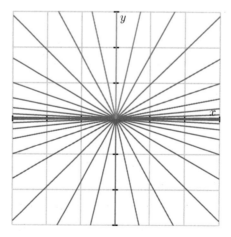

Figure 10.6: Contour lines of $f(x,y) = \frac{y}{\sqrt{x^2+y^2}}$.

10.1. LIMITS

As we have seen in Activity 10.1, if we approach (a,b) along two different paths and find that $f(x,y)$ has two different limits, we can conclude that $\lim_{(x,y)\to(a,b)} f(x,y)$ does not exist. This is similar to the one-variable example $f(x) = x/|x|$ as shown in 10.7; $\lim_{x\to 0} f(x)$ does not exist because we see different limits as x approaches 0 from the left and the right.

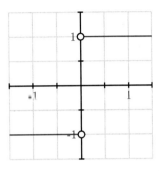

Figure 10.7: The graph of $g(x) = \frac{x}{|x|}$.

As a general rule, we have

If $f(x,y)$ has two different limits as (x,y) approaches (a,b) along two different paths, then $\lim_{(x,y)\to(a,b)} f(x,y)$ does not exist.

As the next activity shows, studying the limit of a two-variable function f by considering the behavior of f along various paths can require subtle insights.

Activity 10.2.

Let's consider the function g defined by

$$g(x,y) = \frac{x^2 y}{x^4 + y^2}$$

and investigate the limit $\lim_{(x,y)\to(0,0)} g(x,y)$.

(a) What is the behavior of g on the x-axis? That is, what is $g(x,0)$ and what is the limit of g as (x,y) approaches $(0,0)$ along the x-axis?

(b) What is the behavior of g on the y-axis? That is, what is $g(0,y)$ and what is the limit of g as (x,y) approaches $(0,0)$ along the y-axis?

(c) What is the behavior of g on the line $y = mx$? That is, what is $g(x,mx)$ and what is the limit of g as (x,y) approaches $(0,0)$ along the line $y = mx$?

(d) Based on what you have seen so far, do you think $\lim_{(x,y)\to(0,0)} g(x,y)$ exists? If so, what do you think its value is?

(e) Now consider the behavior of g on the parabola $y = x^2$? What is $g(x,x^2)$ and what is the limit of g as (x,y) approaches $(0,0)$ along this parabola?

(f) State whether the limit $\lim_{(x,y)\to(0,0)} g(x,y)$ exists or not and provide a justification of your statement.

◁

This activity shows that we need to be careful when studying the limit of a two-variable functions by considering its behavior along different paths. If we find two different paths that result in two different limits, then we may conclude that the limit does not exist. However, we can never conclude that the limit of a function exists only by considering its behavior along different paths.

Generally speaking, concluding that a limit $\lim_{(x,y)\to(a,b)} f(x,y)$ exists requires a more careful argument. For example, let's consider the function f defined by

$$f(x,y) = \frac{x^2 y^2}{x^2 + y^2}$$

and ask whether $\lim_{(x,y)\to(0,0)} f(x,y)$ exists.

Note that if either x or y is 0, then $f(x,y) = 0$. Therefore, if f has a limit at $(0,0)$, it must be 0. We will therefore argue that

$$\lim_{(x,y)\to(0,0)} f(x,y) = 0,$$

by showing that we can make $f(x,y)$ as close to 0 as we wish by taking (x,y) sufficiently close (but not equal) to $(0,0)$. In what follows, we view x and y as being real numbers that are close, but not equal, to 0.

Since $0 \leq x^2$, we have

$$y^2 \leq x^2 + y^2,$$

which implies that

$$\frac{y^2}{x^2 + y^2} \leq 1.$$

Multiplying both sides by x^2 and observing that $f(x,y) \geq 0$ for all (x,y) gives

$$0 \leq f(x,y) = \frac{x^2 y^2}{x^2 + y^2} = x^2 \left(\frac{y^2}{x^2 + y^2} \right) \leq x^2.$$

This shows that we can make $f(x,y)$ as close to 0 as we like by taking x sufficiently close to 0 (for this example, it turns out that we don't even need to worry about making y close to 0). Therefore,

$$\lim_{(x,y)\to(0,0)} \frac{x^2 y^2}{x^2 + y^2} = 0.$$

In spite of the fact that these two most recent examples illustrate some of the complications that arise when studying limits of two-variable functions, many of the properties that are familiar

10.1. LIMITS

from our study of single variable functions hold in precisely the same way. For instance,

> **Properties of Limits.** Let $f = f(x,y)$ and $g = g(x,y)$ be functions so that $\lim_{(x,y)\to(a,b)} f(x,y)$ and $\lim_{(x,y)\to(a,b)} g(x,y)$ both exist. Then
>
> 1. $\lim_{(x,y)\to(a,b)} x = a$ and $\lim_{(x,y)\to(a,b)} y = b$.
>
> 2. $\lim_{(x,y)\to(a,b)} cf(x,y) = c\left(\lim_{(x,y)\to(a,b)} f(x,y)\right)$ for any scalar c,
>
> 3. $\lim_{(x,y)\to(a,b)} [f(x,y) \pm g(x,y)] = \lim_{(x,y)\to(a,b)} f(x,y) \pm \lim_{(x,y)\to(a,b)} g(x,y)$,
>
> 4. $\lim_{(x,y)\to(a,b)} [f(x,y) \cdot g(x,y)] = \left(\lim_{(x,y)\to(a,b)} f(x,y)\right) \cdot \left(\lim_{(x,y)\to(a,b)} g(x,y)\right)$,
>
> 5. $\lim_{(x,y)\to(a,b)} \frac{f(x,y)}{g(x,y)} = \frac{\lim_{(x,y)\to(a,b)} f(x,y)}{\lim_{(x,y)\to(a,b)} g(x,y)}$ if $\lim_{(x,y)\to(a,b)} g(x,y) \neq 0$.

We can use these properties and results from single variable calculus to verify that many limits exist. For example, these properties show that the function f defined by

$$f(x,y) = 3x^2y^3 + 2xy^2 - 3x + 1$$

has a limit at every point (a,b) and, moreover,

$$\lim_{(x,y)\to(a,b)} f(x,y) = f(a,b).$$

The reason for this is that polynomial functions of a single variable have limits at every point.

Continuity

Recall that a function f of a single variable x is said to be continuous at $x = a$ provided that the following three conditions are satisfied:

1. $f(a)$ exists,

2. $\lim_{x \to a} f(x)$ exists, and

3. $\lim_{x \to a} f(x) = f(a)$.

Using our understanding of limits of multivariable functions, we can define continuity in the same way.

> **Definition 10.2.** A function $f = f(x, y)$ is **continuous** at the point (a, b) provided that
>
> 1. f is defined at the point (a, b),
>
> 2. $\lim_{(x,y) \to (a,b)} f(x, y)$ exists, and
>
> 3. $\lim_{(x,y) \to (a,b)} f(x, y) = f(a, b)$.

For instance, we have seen that the function f defined by $f(x, y) = 3x^2y^3 + 2xy^2 - 3x + 1$ is continuous at every point. And just as with single variable functions, continuity has certain properties that are based on the properties of limits.

> **Properties of Continuity.** Let f and g be functions of two variables that are continuous at the point (a, b). Then
>
> 1. cf is continuous at (a, b) for any scalar c
>
> 2. $f + g$ is continuous at (a, b)
>
> 3. $f - g$ is continuous at (a, b)
>
> 4. fg is continuous at (a, b)
>
> 5. $\frac{f}{g}$ is continuous at (a, b) if $g(a, b) \neq 0$

Using these properties, we can apply results from single variable calculus to decide about continuity of multivariable functions. For example, the coordinate functions f and g defined by $f(x, y) = x$ and $g(x, y) = y$ are continuous at every point. We can then use properties of continuity listed to conclude that every polynomial function in x and y is continuous at every point. For example, $g(x, y) = x^2$ and $h(x, y) = y^3$ are continuous functions, so their product $f(x, y) = x^2y^3$ is a continuous multivariable function.

Summary

- A function $f = f(x, y)$ has a limit L at a point (a, b) provided that we can make $f(x, y)$ as close to L as we like by taking (x, y) sufficiently close (but not equal) to (a, b).

- If (x, y) has two different limits as (x, y) approaches (a, b) along two different paths, we can conclude that $\lim_{(x,y) \to (a,b)} f(x, y)$ does not exist.

- Properties similar to those for one-variable functions allow us to conclude that many limits exist and to evaluate them.

10.1. LIMITS

- A function $f = f(x,y)$ is continuous at a point (a,b) in its domain if f has a limit at (a,b) and

$$f(a,b) = \lim_{(x,y)\to(a,b)} f(x,y).$$

Exercises

1. Consider the function f defined by $f(x,y) = \dfrac{xy}{x^2+y^2+1}$.

   (a) What is the domain of f?

   (b) Evaluate limit of f at $(0,0)$ along the following paths: $x=0$, $y=0$, $y=x$, and $y=x^2$.

   (c) What do you conjecture is the value of $\lim_{(x,y)\to(0,0)} f(x,y)$?

   (d) Is f continuous at $(0,0)$? Why or why not?

   (e) Use appropriate technology to sketch both surface and contour plots of f near $(0,0)$. Write several sentences to say how your plots affirm your findings in (a) - (d).

2. Consider the function g defined by $g(x,y) = \dfrac{xy}{x^2+y^2}$.

   (a) What is the domain of g?

   (b) Evaluate limit of g at $(0,0)$ along the following paths: $x=0$, $y=x$, and $y=2x$.

   (c) What can you now say about the value of $\lim_{(x,y)\to(0,0)} g(x,y)$?

   (d) Is g continuous at $(0,0)$? Why or why not?

   (e) Use appropriate technology to sketch both surface and contour plots of g near $(0,0)$. Write several sentences to say how your plots affirm your findings in (a) - (d).

3. For each of the following prompts, provide an example of a function of two variables with the desired properties (with justification), or explain why such a function does not exist.

   (a) A function p that is defined at $(0,0)$, but $\lim_{(x,y)\to(0,0)} p(x,y)$ does not exist.

   (b) A function q that does not have a limit at $(0,0)$, but that has the same limiting value along any line $y = mx$ as $x \to 0$.

   (c) A function r that is continuous at $(0,0)$, but $\lim_{(x,y)\to(0,0)} r(x,y)$ does not exist.

   (d) A function s such that

   $$\lim_{(x,x)\to(0,0)} s(x,x) = 3 \text{ and } \lim_{(x,2x)\to(0,0)} s(x,2x) = 6,$$

   for which $\lim_{(x,y)\to(0,0)} s(x,y)$ exists.

(e) A function t that is not defined at $(1,1)$ but $\lim\limits_{(x,y)\to(1,1)} t(x,y)$ does exist.

10.2 First-Order Partial Derivatives

Motivating Questions

In this section, we strive to understand the ideas generated by the following important questions:

- How are the first-order partial derivatives of a function f of the independent variables x and y defined?

- Given a function f of the independent variables x and y, what do the first-order partial derivatives $\frac{\partial f}{\partial x}$ and $\frac{\partial f}{\partial y}$ tell us about f?

Introduction

The derivative plays a central role in first semester calculus because it provides important information about a function. Thinking graphically, for instance, the derivative at a point tells us the slope of the tangent line to the graph at that point. In addition, the derivative at a point also provides the instantaneous rate of change of the function with respect to changes in the independent variable.

Now that we are investigating functions of two or more variables, we can still ask how fast the function is changing, though we have to be careful about what we mean. Thinking graphically again, we can try to measure how steep the graph of the function is in a particular direction. Alternatively, we may want to know how fast a function's output changes in response to a change in one of the inputs. Over the next few sections, we will develop tools for addressing issues such as these these. Preview Activity 10.2 explores some issues with what we will come to call *partial derivatives*.

Preview Activity 10.2. Let's return to the function we considered in Preview Activity 9.1. Suppose we take out a $18,000 car loan at interest rate r and we agree to pay off the loan in t years. The monthly payment, in dollars, is

$$M(r,t) = \frac{1500r}{1 - \left(1 + \frac{r}{12}\right)^{-12t}}.$$

(a) What is the monthly payment if the interest rate is 3% so that $r = 0.03$, and we pay the loan off in $t = 4$ years?

(b) Suppose the interest rate is fixed at 3%. Express M as a function f of t alone using $r = 0.03$. That is, let $f(t) = M(0.03, t)$. Sketch the graph of f on the left of Figure 10.8. Explain the meaning of the function f.

(c) Find the instantaneous rate of change $f'(4)$ and state the units on this quantity. What information does $f'(4)$ tell us about our car loan? What information does $f'(4)$ tell us about the graph you sketched in (b)?

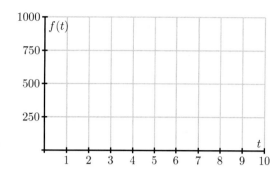
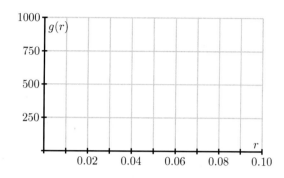

Figure 10.8: The graphs of $f(t) = M(0.03, t)$ and $g(r) = M(r, 4)$.

(d) Express M as a function of r alone, using a fixed time of $t = 4$. That is, let $g(r) = M(r, 4)$. Sketch the graph of g on the right of Figure 10.8. Explain the meaning of the function g.

(e) Find the instantaneous rate of change $g'(0.03)$ and state the units on this quantity. What information does $g'(0.03)$ tell us about our car loan? What information does $g'(0.03)$ tell us about the graph you sketched in (d)?

⋈

First-Order Partial Derivatives

In Section 9.1, we studied the behavior of a function of two or more variables by considering the *traces* of the function. Recall that in one example, we considered the function f defined by

$$f(x, y) = \frac{x^2 \sin(2y)}{32},$$

which measures the range, or horizontal distance, in feet, traveled by a projectile launched with an initial speed of x feet per second at an angle y radians to the horizontal. The graph of this function is given again on the left in Figure 10.9. Moreover, if we fix the angle $y = 0.6$, we may view the trace $f(x, 0.6)$ as a function of x alone, as seen at right in Figure 10.9.

Since the trace is a one-variable function, we may consider its derivative just as we did in the first semester of calculus. With $y = 0.6$, we have

$$f(x, 0.6) = \frac{\sin(1.2)}{32} x^2,$$

and therefore

$$\frac{d}{dx}[f(x, 0.6)] = \frac{\sin(1.2)}{16} x.$$

When $x = 150$, this gives

$$\frac{d}{dx}[f(x, 0.6)]|_{x=150} = \frac{\sin(1.2)}{16} 150 \approx 8.74 \text{ feet per feet per second},$$

10.2. FIRST-ORDER PARTIAL DERIVATIVES

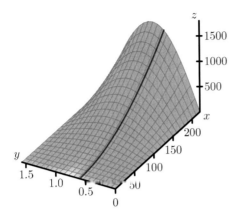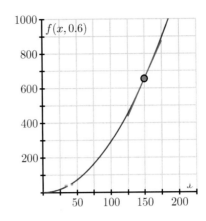

Figure 10.9: The trace with $y = 0.6$.

which gives the slope of the tangent line shown on the right of Figure 10.9. Thinking of this derivative as an instantaneous rate of change implies that if we increase the initial speed of the projectile by one foot per second, we expect the horizontal distance traveled to increase by approximately 8.74 feet if we hold the launch angle constant at 0.6 radians.

By holding y fixed and differentiating with respect to x, we obtain the first-order *partial derivative of f with respect to x*. Denoting this partial derivative as f_x, we have seen that

$$f_x(150, 0.6) = \frac{d}{dx} f(x, 0.6)|_{x=150} = \lim_{h \to 0} \frac{f(150 + h, 0.6) - f(150, 0.6)}{h}.$$

More generally, we have

$$f_x(a, b) = \lim_{h \to 0} \frac{f(a + h, b) - f(a, b)}{h},$$

provided this limit exists.

In the same way, we may obtain a trace by setting, say, $x = 150$ as shown in Figure 10.10. This gives

$$f(150, y) = \frac{150^2}{32} \sin(2y),$$

and therefore

$$\frac{d}{dy}[f(150, y)] = \frac{150^2}{16} \cos(2y).$$

If we evaluate this quantity at $y = 0.6$, we have

$$\frac{d}{dy}[f(150, y)]|_{y=0.6} = \frac{150^2}{16} \cos(1.2) \approx 509.5 \text{ feet per radian}.$$

Once again, the derivative gives the slope of the tangent line shown on the right in Figure 10.10. Thinking of the derivative as an instantaneous rate of change, we expect that the range of the

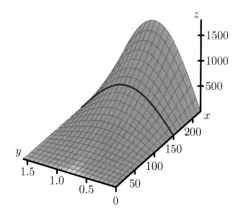
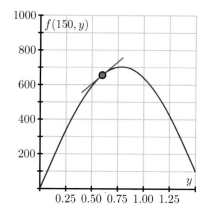

Figure 10.10: The trace with $x = 150$.

projectile increases by 509.5 feet for every radian we increase the launch angle y if we keep the initial speed of the projectile constant at 150 feet per second.

By holding x fixed and differentiating with respect to y, we obtain the first-order *partial derivative of f with respect to y*. As before, we denote this partial derivative as f_y and write

$$f_y(150, 0.6) = \frac{d}{dy} f(150, y)|_{y=0.6} = \lim_{h \to 0} \frac{f(150, 0.6 + h) - f(150, 0.6)}{h}.$$

As with the partial derivative with respect to x, we may express this quantity more generally at an arbitrary point (a, b). To recap, we have now arrived at the formal definition of the first-order partial derivatives of a function of two variables.

Definition 10.3. The first-order **partial derivatives of f with respect to x and y at a point (a, b)** are, respectively,

$$f_x(a, b) = \lim_{h \to 0} \frac{f(a + h, b) - f(a, b)}{h}, \text{ and}$$

$$f_y(a, b) = \lim_{h \to 0} \frac{f(a, b + h) - f(a, b)}{h},$$

provided the limits exist.

Activity 10.3.

Consider the function f defined by

$$f(x, y) = \frac{xy^2}{x + 1}$$

10.2. FIRST-ORDER PARTIAL DERIVATIVES

at the point $(1, 2)$.

(a) Write the trace $f(x, 2)$ at the fixed value $y = 2$. On the left side of Figure 10.11, draw the graph of the trace with $y = 2$ indicating the scale and labels on the axes. Also, sketch the tangent line at the point $x = 1$.

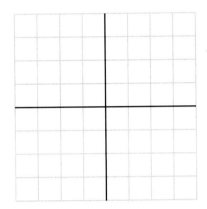 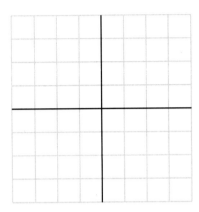

Figure 10.11: Traces of $f(x, y) = \frac{xy^2}{x+1}$.

(b) Find the partial derivative $f_x(1, 2)$ and relate its value to the sketch you just made.

(c) Write the trace $f(1, y)$ at the fixed value $x = 1$. On the right side of Figure 10.11, draw the graph of the trace with $x = 1$ indicating the scale and labels on the axes. Also, sketch the tangent line at the point $y = 2$.

(d) Find the partial derivative $f_y(1, 2)$ and relate its value to the sketch you just made.

◁

As these examples show, each partial derivative at a point arises as the derivative of a one-variable function defined by fixing one of the coordinates. In addition, we may consider each partial derivative as defining a new function of the point (x, y), just as the derivative $f'(x)$ defines a new function of x in single-variable calculus. Due to the connection between one-variable derivatives and partial derivatives, we will often use Leibniz-style notation to denote partial derivatives by writing

$$\frac{\partial f}{\partial x}(a, b) = f_x(a, b), \quad \text{and} \quad \frac{\partial f}{\partial y}(a, b) = f_y(a, b).$$

To calculate the partial derivative f_x, we hold y fixed and thus we treat y as a constant. In Leibniz notation, observe that

$$\frac{\partial}{\partial x}(x) = 1 \quad \text{and} \quad \frac{\partial}{\partial x}(y) = 0.$$

To see the contrast between how we calculate single variable derivatives and partial derivatives, and the difference between the notations $\frac{d}{dx}[\ ]$ and $\frac{\partial}{\partial x}[\ ]$, observe that

$$\frac{d}{dx}[3x^2 - 2x + 3] = 3\frac{d}{dx}[x^2] - 2\frac{d}{dx}[x] + \frac{d}{dx}[3] = 3 \cdot 2x - 2,$$

and $\quad \dfrac{\partial}{\partial x}[x^2 y - xy + 2y] = y\dfrac{\partial}{\partial x}[x^2] - y\dfrac{\partial}{\partial x}[x] + \dfrac{\partial}{\partial x}[2y] = y \cdot 2x - y$

Thus, computing partial derivatives is straightforward: we use the standard rules of single variable calculus, but do so while holding one (or more) of the variables constant.

Activity 10.4.

(a) If we have the function f of the variables x and y and we want to find the partial derivative f_x, which variable do we treat as a constant? When we find the partial derivative f_y, which variable do we treat as a constant?

(b) If $f(x, y) = 3x^3 - 2x^2 y^5$, find the partial derivatives f_x and f_y.

(c) If $f(x, y) = \dfrac{xy^2}{x+1}$, find the partial derivatives f_x and f_y.

(d) If $g(r, s) = rs \cos(r)$, find the partial derivatives g_r and g_s.

(e) Assuming $f(w, x, y) = (6w + 1)\cos(3x^2 + 4xy^3 + y)$, find the partial derivatives f_w, f_x, and f_y.

(f) Find all possible first-order partial derivatives of $q(x, t, z) = \dfrac{x2^t z^3}{1 + x^2}$.

◁

Interpretations of First-Order Partial Derivatives

Recall that the derivative of a single variable function has a geometric interpretation as the slope of the line tangent to the graph at a given point. Similarly, we have seen that the partial derivatives measure the slope of a line tangent to a trace of a function of two variables as shown in Figure 10.12.

Now we consider the first-order partial derivatives in context. Recall that the difference quotient $\frac{f(a+h)-f(a)}{h}$ for a function f of a single variable x at a point where $x = a$ tells us the average rate of change of f over the interval $[a, a + h]$, while the derivative $f'(a)$ tells us the instantaneous rate of change of f at $x = a$. We can use these same concepts to explain the meanings of the partial derivatives in context.

Activity 10.5.

The speed of sound C traveling through ocean water is a function of temperature, salinity and depth. It may be modeled by the function

$$C = 1449.2 + 4.6T - 0.055T^2 + 0.00029T^3 + (1.34 - 0.01T)(S - 35) + 0.016D.$$

10.2. FIRST-ORDER PARTIAL DERIVATIVES

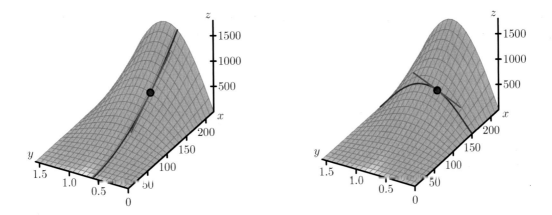

Figure 10.12: Tangent lines to two traces of the range function.

Here C is the speed of sound in meters/second, T is the temperature in degrees Celsius, S is the salinity in grams/liter of water, and D is the depth below the ocean surface in meters.

(a) State the units in which each of the partial derivatives, C_T, C_S and C_D, are expressed and explain the physical meaning of each.

(b) Find the partial derivatives C_T, C_S and C_D.

(c) Evaluate each of the three partial derivatives at the point where $T = 10$, $S = 35$ and $D = 100$. What does the sign of each partial derivatives tell us about the behavior of the function C at the point $(10, 35, 100)$?

◁

Using tables and contours to estimate partial derivatives

Remember that functions of two variables are often represented as either a table of data or a contour plot. In single variable calculus, we saw how we can use the difference quotient to approximate derivatives if, instead of an algebraic formula, we only know the value of the function at a few points. The same idea applies to partial derivatives.

Activity 10.6.

The wind chill, as frequently reported, is a measure of how cold it feels outside when the wind is blowing. In Table 10.1, the wind chill w, measured in degrees Fahrenheit, is a function of the wind speed v, measured in miles per hour, and the ambient air temperature T, also measured in degrees Fahrenheit. We thus view w as being of the form $w = w(v, T)$.

(a) Estimate the partial derivative $w_v(20, -10)$. What are the units on this quantity and what does it mean?

v\T	-30	-25	-20	-15	-10	-5	0	5	10	15	20
5	-46	-40	-34	-28	-22	-16	-11	-5	1	7	13
10	-53	-47	-41	-35	-28	-22	-16	-10	-4	3	9
15	-58	-51	-45	-39	-32	-26	-19	-13	-7	0	6
20	-61	-55	-48	-42	-35	-29	-22	-15	-9	-2	4
25	-64	-58	-51	-44	-37	-31	-24	-17	-11	-4	3
30	-67	-60	-53	-46	-39	-33	-26	-19	-12	-5	1
35	-69	-62	-55	-48	-41	-34	-27	-21	-14	-7	0
40	-71	-64	-57	-50	-43	-36	-29	-22	-15	-8	-1

Table 10.1: Wind chill as a function of wind speed and temperature.

(b) Estimate the partial derivative $w_T(20, -10)$. What are the units on this quantity and what does it mean?

(c) Use your results to estimate the wind chill $w(18, -10)$.

(d) Use your results to estimate the wind chill $w(20, -12)$.

(e) Use your results to estimate the wind chill $w(18, -12)$.

◁

Activity 10.7.

Shown below in Figure 10.13 is a contour plot of a function f. The value of the function along a few of the contours is indicated to the left of the figure.

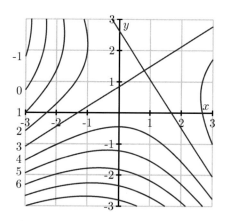

Figure 10.13: A contour plot of f.

(a) Estimate the partial derivative $f_x(-2, -1)$.

(b) Estimate the partial derivative $f_y(-2, -1)$.

(c) Estimate the partial derivatives $f_x(-1, 2)$ and $f_y(-1, 2)$.

10.2. FIRST-ORDER PARTIAL DERIVATIVES

(d) Locate one point (x, y) where the partial derivative $f_x(x, y) = 0$.

(e) Locate one point (x, y) where $f_x(x, y) < 0$.

(f) Locate one point (x, y) where $f_y(x, y) > 0$.

(g) Suppose you have a different function g, and you know that $g(2, 2) = 4$, $g_x(2, 2) > 0$, and $g_y(2, 2) > 0$. Using this information, sketch a possibility for the contour $g(x, y) = 4$ passing through $(2, 2)$ on the left side of Figure 10.14. Then include possible contours $g(x, y) = 3$ and $g(x, y) = 5$.

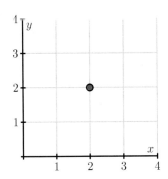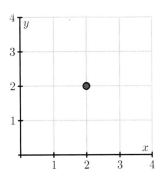

Figure 10.14: Plots for contours of g and h.

(h) Suppose you have yet another function h, and you know that $h(2, 2) = 4$, $h_x(2, 2) < 0$, and $h_y(2, 2) > 0$. Using this information, sketch a possible contour $h(x, y) = 4$ passing through $(2, 2)$ on the right side of Figure 10.14. Then include possible contours $h(x, y) = 3$ and $h(x, y) = 5$.

◁

Summary

- If $f = f(x, y)$ is a function of two variables, there are two first order partial derivatives of f: the partial derivative of f with respect to x,

$$\frac{\partial f}{\partial x}(x, y) = f_x(x, y) = \lim_{h \to 0} \frac{f(x + h, y) - f(x, y)}{h},$$

and the partial derivative of f with respect to y,

$$\frac{\partial f}{\partial y}(x, y) = f_y(x, y) = \lim_{h \to 0} \frac{f(x, y + h) - f(x, y)}{h},$$

where each partial derivative exists only at those points (x, y) for which the limit exists.

- The partial derivative $f_x(a, b)$ tells us the instantaneous rate of change of f with respect to x at $(x, y) = (a, b)$ when y is fixed at b. Geometrically, the partial derivative $f_x(a, b)$ tells us the slope of the line tangent to the $y = b$ trace of the function f at the point $(a, b, f(a, b))$.

- The partial derivative $f_y(a,b)$ tells us the instantaneous rate of change of f with respect to y at $(x,y) = (a,b)$ when x is fixed at a. Geometrically, the partial derivative $f_y(a,b)$ tells us the slope of the line tangent to the $x = a$ trace of the function f at the point $(a, b, f(a,b))$.

Exercises

1. The Heat Index, I, (measured in *apparent degrees* F) is a function of the actual temperature T outside (in degrees F) and the relative humidity H (measured as a percentage). A portion of the table which gives values for this function, $I = I(T, H)$, is reproduced below:

$T \downarrow \setminus H \rightarrow$	70	75	80	85
90	106	109	112	115
92	112	115	119	123
94	118	122	127	132
96	125	130	135	141

   (a) State the limit definition of the value $I_T(94, 75)$. Then, estimate $I_T(94, 75)$, and write one complete sentence that carefully explains the meaning of this value, including its units.

   (b) State the limit definition of the value $I_H(94, 75)$. Then, estimate $I_H(94, 75)$, and write one complete sentence that carefully explains the meaning of this value, including its units.

   (c) Suppose you are given that $I_T(92, 80) = 3.75$ and $I_H(92, 80) = 0.8$. Estimate the values of $I(91, 80)$ and $I(92, 78)$. Explain how the partial derivatives are relevant to your thinking.

   (d) On a certain day, at 1 p.m. the temperature is 92 degrees and the relative humidity is 85%. At 3 p.m., the temperature is 96 degrees and the relative humidity 75%. What is the average rate of change of the heat index over this time period, and what are the units on your answer? Write a sentence to explain your thinking.

2. Let $f(x, y) = \frac{1}{2}xy^2$ represent the kinetic energy in Joules of an object of mass x in kilograms with velocity y in meters per second. Let (a, b) be the point $(4, 5)$ in the domain of f.

   (a) Calculate $f_x(a, b)$.

   (b) Explain as best you can in the context of kinetic energy what the partial derivative

   $$f_x(a, b) = \lim_{h \to 0} \frac{f(a+h, b) - f(a, b)}{h}$$

   tells us about kinetic energy.

   (c) Calculate $f_y(a, b)$.

(d) Explain as best you can in the context of kinetic energy what the partial derivative

$$f_y(a,b) = \lim_{h \to 0} \frac{f(a, b+h) - f(a,b)}{h}$$

tells us about kinetic energy.

(e) Often we are given certain graphical information about a function instead of a rule. We can use that information to approximate partial derivatives. For example, suppose that we are given a contour plot of the kinetic energy function (as in Figure 10.15) instead of a formula. Use this contour plot to approximate $f_x(4,5)$ and $f_y(4,5)$ as best you can. Compare to your calculations from earlier parts of this exercise.

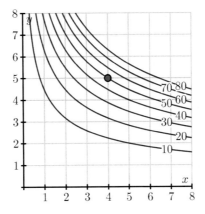

Figure 10.15: The graph of $f(x,y) = \frac{1}{2}xy^2$.

3. The temperature on an unevenly heated metal plate positioned in the first quadrant of the x-y plane is given by

$$C(x,y) = \frac{25xy + 25}{(x-1)^2 + (y-1)^2 + 1}.$$

Assume that temperature is measured in degrees Celsius and that x and y are each measured in inches. (**Note:** At no point in the following questions should you expand the denominator of $C(x,y)$.)

(a) Determine $\frac{\partial C}{\partial x}|_{(x,y)}$ and $\frac{\partial C}{\partial y}|_{(x,y)}$.

(b) If an ant is on the metal plate, standing at the point $(2,3)$, and starts walking in a direction parallel to the y axis, at what rate will the temperature he is experiencing change? Explain, and include appropriate units.

(c) If an ant is walking along the line $y = 3$, at what instantaneous rate will the temperature he is experiencing change when he passes the point $(1,3)$?

(d) Now suppose the ant is stationed at the point $(6,3)$ and walks in a straight line towards the point $(2,0)$. Determine the *average* rate of change in temperature (per unit distance traveled) the ant encounters in moving between these two points. Explain your reasoning carefully. What are the units on your answer?

4. Consider the function f defined by $f(x,y) = 8 - x^2 - 3y^2$.

   (a) Determine $f_x(x,y)$ and $f_y(x,y)$.

   (b) Find parametric equations in $\mathbb{R}^3$ for the tangent line to the trace $f(x,1)$ at $x=2$.

   (c) Find parametric equations in $\mathbb{R}^3$ for the tangent line to the trace $f(2,y)$ at $y=1$.

   (d) State respective direction vectors for the two lines determined in (b) and (c).

   (e) Determine the equation of the plane that passes through the point $(2,1,f(2,1))$ whose normal vector is orthogonal to the direction vectors of the two lines found in (b) and (c).

   (f) Use a graphing utility to plot both the surface $z = 8 - x^2 - 3y^2$ and the plane from (e) near the point $(2,1)$. What is the relationship between the surface and the plane?

10.3 Second-Order Partial Derivatives

Motivating Questions

In this section, we strive to understand the ideas generated by the following important questions:

- Given a function f of two independent variables x and y, how are the second-order partial derivatives of f defined?

- What do the second-order partial derivatives f_{xx}, f_{yy}, f_{xy}, and f_{yx} of a function f tell us about the function's behavior?

Introduction

Recall that for a single-variable function f, the second derivative of f is defined to be the derivative of the first derivative. That is, $f''(x) = \frac{d}{dx}[f'(x)]$, which can be stated in terms of the limit definition of the derivative by writing

$$f''(x) = \lim_{h \to 0} \frac{f'(x+h) - f'(x)}{h}.$$

In what follows, we begin exploring the four different second-order partial derivatives of a function of two variables and seek to understand what these various derivatives tell us about the function's behavior.

Preview Activity 10.3. Once again, let's consider the function f defined by $f(x, y) = \frac{x^2 \sin(2y)}{32}$ that

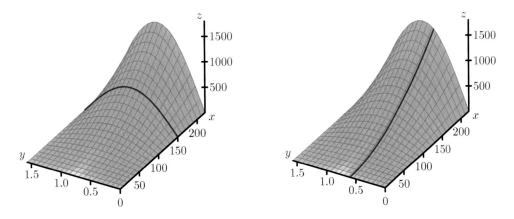

Figure 10.16: The range function with traces $y = 0.6$ and $x = 150$.

measures a projectile's range as a function of its initial speed x and launch angle y. The graph of this function, including traces with $x = 150$ and $y = 0.6$, is shown in Figure 10.16.

(a) Compute the partial derivative f_x and notice that f_x itself is a new function of x and y.

(b) We may now compute the partial derivatives of f_x. Find the partial derivative $f_{xx} = (f_x)_x$ and evaluate $f_{xx}(150, 0.6)$.

(c) Figure 10.17 shows the trace of f with $y = 0.6$ with three tangent lines included. Explain how your result from part (b) of this preview activity is reflected in this figure.

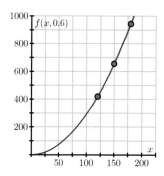

Figure 10.17: The trace with $y = 0.6$.

(d) Determine the partial derivative f_y, and then find the partial derivative $f_{yy} = (f_y)_y$. Evaluate $f_{yy}(150, 0.6)$.

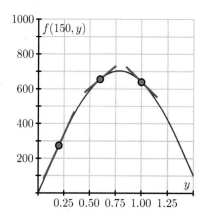

Figure 10.18: More traces of the range function.

(e) Figure 10.18 shows the trace $f(150, y)$ and includes three tangent lines. Explain how the value of $f_{yy}(150, 0.6)$ is reflected in this figure.

(f) Because f_x and f_y are each functions of both x and y, they each have two partial derivatives. Not only can we compute $f_{xx} = (f_x)_x$, but also $f_{xy} = (f_x)_y$; likewise, in addition to $f_{yy} = (f_y)_y$, but also $f_{yx} = (f_y)_x$. For the range function $f(x, y) = \frac{x^2 \sin(2y)}{32}$, use your earlier computations of f_x and f_y to now determine f_{xy} and f_{yx}. Write one sentence to explain how you calculated these "mixed" partial derivatives.

Second-order Partial Derivatives

A function f of two independent variables x and y has two first order partial derivatives, f_x and f_y. As we saw in Preview Activity 10.3, each of these first-order partial derivatives has two partial derivatives, giving a total of four *second-order* partial derivatives:

- $f_{xx} = (f_x)_x = \frac{\partial}{\partial x}\left(\frac{\partial f}{\partial x}\right) = \frac{\partial^2 f}{\partial x^2}$,
- $f_{yy} = (f_y)_y = \frac{\partial}{\partial y}\left(\frac{\partial f}{\partial y}\right) = \frac{\partial^2 f}{\partial y^2}$,
- $f_{xy} = (f_x)_y = \frac{\partial}{\partial y}\left(\frac{\partial f}{\partial x}\right) = \frac{\partial^2 f}{\partial y \partial x}$,
- $f_{yx} = (f_y)_x = \frac{\partial}{\partial x}\left(\frac{\partial f}{\partial y}\right) = \frac{\partial^2 f}{\partial x \partial y}$.

The first two are called *unmixed* second-order partial derivatives while the last two are called the *mixed* second-order partial derivatives.

One aspect of this notation can be a little confusing. The notation

$$\frac{\partial^2 f}{\partial y \partial x} = \frac{\partial}{\partial y}\left(\frac{\partial f}{\partial x}\right)$$

means that we first differentiate with respect to x and then with respect to y; this can be expressed in the alternate notation $f_{xy} = (f_x)_y$. However, to find the second partial derivative

$$f_{yx} = (f_y)_x$$

we first differentiate with respect to y and then x. This means that

$$\frac{\partial^2 f}{\partial y \partial x} = f_{xy}, \quad \text{and} \quad \frac{\partial^2 f}{\partial x \partial y} = f_{yx}.$$

Be sure to note carefully the difference between Leibniz notation and subscript notation and the order in which x and y appear in each. In addition, remember that anytime we compute a partial derivative, we hold constant the variable(s) other than the one we are differentiating with respect to.

Activity 10.8.

Find all second order partial derivatives of the following functions. For each partial derivative you calculate, state explicitly which variable is being held constant.

(a) $f(x, y) = x^2 y^3$

(b) $f(x, y) = y \cos(x)$

(c) $g(s,t) = st^3 + s^4$

(d) How many second order partial derivatives does the function h defined by $h(x,y,z) = 9x^9z - xyz^9 + 9$ have? Find h_{xz} and h_{zx}.

◁

In Preview Activity 10.3 and Activity 10.8, you may have noticed that the mixed second-order partial derivatives are equal. This observation holds generally and is known as Clairaut's Theorem.

> **Clairaut's Theorem.** Let f be a function of two variables for which the partial derivatives f_{xy} and f_{yx} are continuous near the point (a,b). Then
>
> $$f_{xy}(a,b) = f_{yx}(a,b).$$

Interpreting the second-order Partial Derivatives

Recall from single variable calculus that the second derivative measures the instantaneous rate of change of the derivative. This observation is the key to understanding the meaning of the second-order partial derivatives.

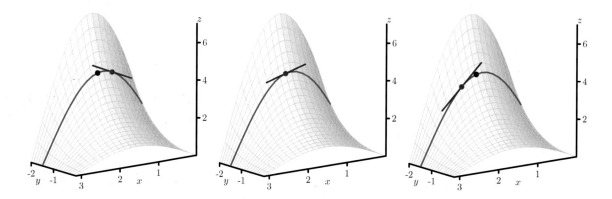

Figure 10.19: The tangent lines to a trace with increasing x.

Furthermore, we remember that the second derivative of a function at a point provides us with information about the concavity of the function at that point. Since the unmixed second-order partial derivative f_{xx} requires us to hold y constant and differentiate twice with respect to x, we may simply view f_{xx} as the second derivative of a trace of f where y is fixed. As such, f_{xx} will measure the concavity of this trace.

Consider, for example, $f(x,y) = \sin(x)e^{-y}$. Figure 10.19 shows the graph of this function along with the trace given by $y = -1.5$. Also shown are three tangent lines to this trace, with increasing

10.3. SECOND-ORDER PARTIAL DERIVATIVES

x-values from left to right among the three plots in Figure 10.19.

That the slope of the tangent line is decreasing as x increases is reflected, as it is in one-variable calculus, in the fact that the trace is concave down. Indeed, we see that $f_x(x, y) = \cos(x)e^{-y}$ and so $f_{xx}(x, y) = -\sin(x)e^{-y} < 0$, since $e^{-y} > 0$ for all values of y, including $y = -1.5$.

In the following activity, we further explore what second-order partial derivatives tell us about the geometric behavior of a surface.

Activity 10.9.

We continue to consider the function f defined by $f(x, y) = \sin(x)e^{-y}$.

(a) In Figure 10.20, we see the trace of $f(x, y) = \sin(x)e^{-y}$ that has x held constant with $x = 1.75$. Write a couple of sentences that describe whether the slope of the tangent

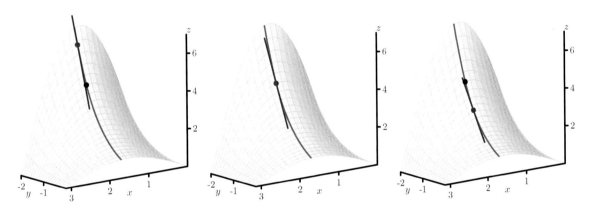

Figure 10.20: The tangent lines to a trace with increasing y.

lines to this curve increase or decrease as y increases, and, after computing $f_{yy}(x, y)$, explain how this observation is related to the value of $f_{yy}(1.75, y)$. Be sure to address the notion of concavity in your response.

(b) In Figure 10.21, we start to think about the mixed partial derivative, f_{xy}. Here, we first hold y constant to generate the first-order partial derivative f_x, and then we hold x constant to compute f_{xy}. This leads to first thinking about a trace with x being constant, followed by slopes of tangent lines in the y-direction that slide along the original trace. You might think of sliding your pencil down the trace with x constant in a way that its slope indicates $(f_x)_y$ in order to further animate the three snapshots shown in the figure. Based on Figure 10.21, is $f_{xy}(1.75, -1.5)$ positive or negative? Why?

(c) Determine the formula for $f_{xy}(x, y)$, and hence evaluate $f_{xy}(1.75, -1.5)$. How does this value compare with your observations in (b)?

(d) We know that $f_{xx}(1.75, -1.5)$ measures the concavity of the $y = -1.5$ trace, and that $f_{yy}(1.75, -1.5)$ measures the concavity of the $x = 1.75$ trace. What do you think the quantity $f_{xy}(1.75, -1.5)$ measures?

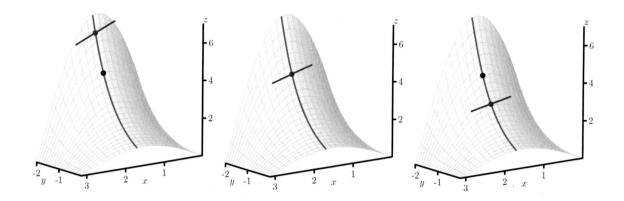

Figure 10.21: The trace of $z = f(x,y) = \sin(x)e^{-y}$ with $x = 1.75$, along with tangent lines in the y-direction at three different points.

(e) On Figure 10.21, sketch the trace with $y = -1.5$, and sketch three tangent lines whose slopes correspond to the value of $f_{yx}(x, -1.5)$ for three different values of x, the middle of which is $x = -1.5$. Is $f_{yx}(1.75, -1.5)$ positive or negative? Why? What does $f_{yx}(1.75, -1.5)$ measure?

◁

Just as with the first-order partial derivatives, we can approximate second-order partial derivatives in the situation where we have only partial information about the function.

Activity 10.10.

As we saw in Activity 10.6, the wind chill $w(v, T)$, in degrees Fahrenheit, is a function of the wind speed, in miles per hour, and the air temperature, in degrees Fahrenheit. Some values of the wind chill are recorded in Table 10.2.

$v \backslash T$	-30	-25	-20	-15	-10	-5	0	5	10	15	20
5	-46	-40	-34	-28	-22	-16	-11	-5	1	7	13
10	-53	-47	-41	-35	-28	-22	-16	-10	-4	3	9
15	-58	-51	-45	-39	-32	-26	-19	-13	-7	0	6
20	-61	-55	-48	-42	-35	-29	-22	-15	-9	-2	4
25	-64	-58	-51	-44	-37	-31	-24	-17	-11	-4	3
30	-67	-60	-53	-46	-39	-33	-26	-19	-12	-5	1
35	-69	-62	-55	-48	-41	-34	-27	-21	-14	-7	0
40	-71	-64	-57	-50	-43	-36	-29	-22	-15	-8	-1

Table 10.2: Wind chill as a function of wind speed and temperature.

(a) Estimate the partial derivatives $w_T(20, -15)$, $w_T(20, -10)$, and $w_T(20, -5)$. Use these results to estimate the second-order partial $w_{TT}(20, -10)$.

10.3. SECOND-ORDER PARTIAL DERIVATIVES

(b) In a similar way, estimate the second-order partial $w_{vv}(20, -10)$.

(c) Estimate the partial derivatives $w_T(20, -10)$, $w_T(25, -10)$, and $w_T(15, -10)$, and use your results to estimate the partial $w_{Tv}(20, -10)$.

(d) In a similar way, estimate the partial derivative $w_{vT}(20, -10)$.

(e) Write several sentences that explain what the values $w_{TT}(20, -10)$, $w_{vv}(20, -10)$, and $w_{Tv}(20, -10)$ indicate regarding the behavior of $w(v, T)$.

◁

As we have found in Activities 10.9 and 10.10, we may think of f_{xy} as measuring the "twist" of the graph as we increase y along a particular trace where x is held constant. In the same way, f_{yx} measures how the graph twists as we increase x. If we remember that Clairaut's theorem tells us that $f_{xy} = f_{yx}$, we see that the amount of twisting is the same in both directions. This twisting is perhaps more easily seen in Figure 10.22, which shows the graph of $f(x, y) = -xy$, for which $f_{xy} = -1$.

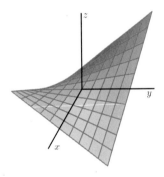

Figure 10.22: The graph of $f(x, y) = -xy$.

Summary

- There are four second-order partial derivatives of a function f of two independent variables x and y:
$$f_{xx} = (f_x)_x, \, f_{xy} = (f_x)_y, \, f_{yx} = (f_y)_x, \text{ and } f_{yy} = (f_y)_y.$$

- The unmixed second-order partial derivatives, f_{xx} and f_{yy}, tell us about the concavity of the traces. The mixed second-order partial derivatives, f_{xy} and f_{yx}, tell us how the graph of f twists.

Exercises

1. Shown in Figure 10.23 is a contour plot of a function f with the values of f labeled on the contours. The point $(2, 1)$ is highlighted in red.

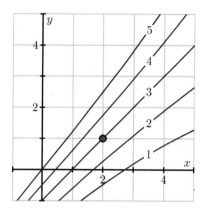

Figure 10.23: A contour plot of $f(x, y)$.

(a) Estimate the partial derivatives $f_x(2, 1)$ and $f_y(2, 1)$.

(b) Determine whether the second-order partial derivative $f_{xx}(2, 1)$ is positive or negative, and explain your thinking.

(c) Determine whether the second-order partial derivative $f_{yy}(2, 1)$ is positive or negative, and explain your thinking.

(d) Determine whether the second-order partial derivative $f_{xy}(2, 1)$ is positive or negative, and explain your thinking.

(e) Determine whether the second-order partial derivative $f_{yx}(2, 1)$ is positive or negative, and explain your thinking.

(f) Consider a function g of the variables x and y for which $g_x(2, 2) > 0$ and $g_{xx}(2, 2) < 0$. Sketch possible behavior of some contours around $(2, 2)$ on the left axes in Figure 10.24.

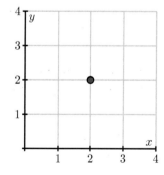 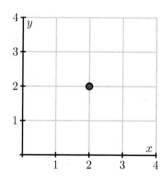

Figure 10.24: Plots for contours of g and h.

(g) Consider a function h of the variables x and y for which $h_x(2, 2) > 0$ and $h_{xy}(2, 2) < 0$.

Sketch possible behavior of some contour lines around $(2,2)$ on the right axes in Figure 10.24.

2. The Heat Index, I, (measured in *apparent degrees* F) is a function of the actual temperature T outside (in degrees F) and the relative humidity H (measured as a percentage). A portion of the table which gives values for this function, $I(T, H)$, is reproduced below:

$T \downarrow \backslash H \rightarrow$	70	75	80	85
90	106	109	112	115
92	112	115	119	123
94	118	122	127	132
96	125	130	135	141

(a) State the limit definition of the value $I_{TT}(94, 75)$. Then, estimate $I_{TT}(94, 75)$, and write one complete sentence that carefully explains the meaning of this value, including units.

(b) State the limit definition of the value $I_{HH}(94, 75)$. Then, estimate $I_{HH}(94, 75)$, and write one complete sentence that carefully explains the meaning of this value, including units.

(c) Finally, do likewise to estimate $I_{HT}(94, 75)$, and write a sentence to explain the meaning of the value you found.

3. The temperature on a heated metal plate positioned in the first quadrant of the x-y plane is given by
$$C(x, y) = 25e^{-(x-1)^2 - (y-1)^3}.$$
Assume that temperature is measured in degrees Celsius and that x and y are each measured in inches.

(a) Determine $C_{xx}(x, y)$ and $C_{yy}(x, y)$. Do not do any additional work to algebraically simplify your results.

(b) Calculate $C_{xx}(1.1, 1.2)$. Suppose that an ant is walking past the point $(1.1, 1.2)$ along the line $y = 1.2$. Write a sentence to explain the meaning of the value of $C_{xx}(1.1, 1.2)$, including units.

(c) Calculate $C_{yy}(1.1, 1.2)$. Suppose instead that an ant is walking past the point $(1.1, 1.2)$ along the line $x = 1.1$. Write a sentence to explain the meaning of the value of $C_{yy}(1.1, 1.2)$, including units.

(d) Determine $C_{xy}(x, y)$ and hence compute $C_{xy}(1.1, 1.2)$. What is the meaning of this value? Explain, in terms of an ant walking on the heated metal plate.

4. Let $f(x, y) = 8 - x^2 - y^2$ and $g(x, y) = 8 - x^2 + 4xy - y^2$.

(a) Determine f_x, f_y, f_{xx}, f_{yy}, f_{xy}, and f_{yx}.

(b) Evaluate each of the partial derivatives in (a) at the point $(0, 0)$.

(c) What do the values in (b) suggest about the behavior of f near $(0,0)$? Plot a graph of f and compare what you see visually to what the values suggest.

(d) Determine g_x, g_y, g_{xx}, g_{yy}, g_{xy}, and g_{yx}.

(e) Evaluate each of the partial derivatives in (d) at the point $(0,0)$.

(f) What do the values in (e) suggest about the behavior of g near $(0,0)$? Plot a graph of g and compare what you see visually to what the values suggest.

(g) What do the functions f and g have in common at $(0,0)$? What is different? What do your observations tell you regarding the importance of a certain second-order partial derivative?

5. Let $f(x,y) = \frac{1}{2}xy^2$ represent the kinetic energy in Joules of an object of mass x in kilograms with velocity y in meters per second. Let (a,b) be the point $(4,5)$ in the domain of f.

(a) Calculate $\dfrac{\partial^2 f}{\partial x^2}$ at the point (a,b). Then explain as best you can what this second order partial derivative tells us about kinetic energy.

(b) Calculate $\dfrac{\partial^2 f}{\partial y^2}$ at the point (a,b). Then explain as best you can what this second order partial derivative tells us about kinetic energy.

(c) Calculate $\dfrac{\partial^2 f}{\partial y \partial x}$ at the point (a,b). Then explain as best you can what this second order partial derivative tells us about kinetic energy.

(d) Calculate $\dfrac{\partial^2 f}{\partial x \partial y}$ at the point (a,b). Then explain as best you can what this second order partial derivative tells us about kinetic energy.

10.4 Linearization: Tangent Planes and Differentials

Motivating Questions

In this section, we strive to understand the ideas generated by the following important questions:

- What does it mean for a function of two variables to be locally linear at a point?
- How do we find the equation of the plane tangent to a locally linear function at a point?
- What does it mean to say that a multivariable function is *differentiable*?
- What is the differential of a multivariable function of two variables and what are its uses?

Introduction

One of the central concepts in single variable calculus is that the graph of a differentiable function, when viewed on a very small scale, looks like a line. We call this line the tangent line and measure its slope with the derivative. In this section, we will extend this concept to functions of several variables.

Let's see what happens when we look at the graph of a two-variable function on a small scale. To begin, let's consider the function f defined by

$$f(x, y) = 6 - \frac{x^2}{2} - y^2,$$

whose graph is shown in Figure 10.25.

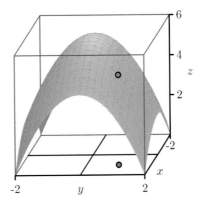

Figure 10.25: The graph of $f(x, y) = 6 - x^2/2 - y^2$.

We choose to study the behavior of this function near the point $(x_0, y_0) = (1, 1)$. In particular, we wish to view the graph on an increasingly small scale around this point, as shown in the two plots in Figure 10.26

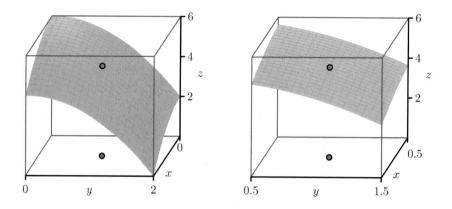

Figure 10.26: The graph of $f(x, y) = 6 - x^2/2 - y^2$.

Just as the graph of a differentiable single-variable function looks like a line when viewed on a small scale, we see that the graph of this particular two-variable function looks like a plane, as seen in Figure 10.27. In the following preview activity, we explore how to find the equation of this plane.

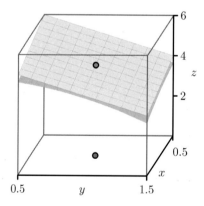

Figure 10.27: The graph of $f(x, y) = 6 - x^2/2 - y^2$.

In what follows, we will also use the important fact[1] that the plane passing through (x_0, y_0, z_0) may be expressed in the form $z = z_0 + a(x - x_0) + b(y - y_0)$, where a and b are constants.

---

[1]As we saw in Section 9.5, the equation of a plane passing through the point (x_0, y_0, z_0) may be written in the form $A(x - x_0) + B(y - y_0) + C(z - z_0) = 0$. If the plane is not vertical, then $C \neq 0$, and we can rearrange this and hence write $C(z - z_0) = -A(x - x_0) - B(y - y_0)$ and thus

$$z = z_0 - \frac{A}{C}(x - x_0) - \frac{B}{C}(y - y_0)$$
$$= z_0 + a(x - x_0) + b(y - y_0)$$

where $a = -A/C$ and $b = -B/C$, respectively.

10.4. LINEARIZATION: TANGENT PLANES AND DIFFERENTIALS

Preview Activity 10.4. Let $f(x,y) = 6 - \frac{x^2}{2} - y^2$, and let $(x_0, y_0) = (1, 1)$.

(a) Evaluate $f(x, y) = 6 - \frac{x^2}{2} - y^2$ and its partial derivatives at (x_0, y_0); that is, find $f(1, 1)$, $f_x(1, 1)$, and $f_y(1, 1)$.

(b) We know one point on the tangent plane; namely, the z-value of the tangent plane agrees with the z-value on the graph of $f(x, y) = 6 - \frac{x^2}{2} - y^2$ at the point (x_0, y_0). In other words, both the tangent plane and the graph of the function f contain the point (x_0, y_0, z_0). Use this observation to determine z_0 in the expression $z = z_0 + a(x - x_0) + b(y - y_0)$.

(c) Sketch the traces of $f(x, y) = 6 - \frac{x^2}{2} - y^2$ for $y = y_0 = 1$ and $x = x_0 = 1$ below in Figure 10.28.

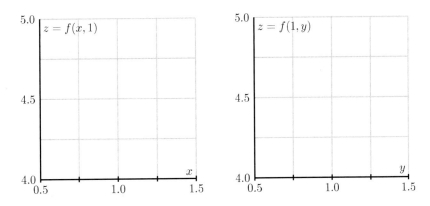

Figure 10.28: The traces of $f(x, y)$ with $y = y_0 = 1$ and $x = x_0 = 1$.

(d) Determine the equation of the tangent line of the trace that you sketched in the previous part with $y = 1$ (in the x direction) at the point $x_0 = 1$.

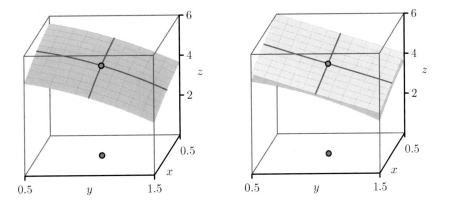

Figure 10.29: The traces of $f(x, y)$ and the tangent plane.

(e) Figure 10.29 shows the traces of the function and the traces of the tangent plane. Explain how the tangent line of the trace of f, whose equation you found in the last part of this activity, is related to the tangent plane. How does this observation help you determine the constant a in the equation for the tangent plane $z = z_0 + a(x - x_0) + b(y - y_0)$? (Hint: How do you think $f_x(x_0, y_0)$ should be related to $z_x(x_0, y_0)$?)

(f) In a similar way to what you did in (d), determine the equation of the tangent line of the trace with $x = 1$ at the point $y_0 = 1$. Explain how this tangent line is related to the tangent plane, and use this observation to determine the constant b in the equation for the tangent plane $z = z_0 + a(x - x_0) + b(y - y_0)$. (Hint: How do you think $f_y(x_0, y_0)$ should be related to $z_y(x_0, y_0)$?)

(g) Finally, write the equation $z = z_0 + a(x - x_0) + b(y - y_0)$ of the tangent plane to the graph of $f(x, y) = 6 - x^2/2 - y^2$ at the point $(x_0, y_0) = (1, 1)$.

◁

The tangent plane

Before stating the formula for the equation of the tangent plane at a point for a general function $f = f(x, y)$, we need to discuss a mild technical condition. As we have noted, when we look at the graph of a single-variable function on a small scale near a point x_0, we expect to see a line; in this case, we say that f is *locally linear near* x_0 since the graph looks like a linear function locally around x_0. Of course, there are functions, such as the absolute value function given by $f(x) = |x|$, that are not locally linear at every point. In single-variable calculus, we learn that if the derivative of a function exists at a point, then the function is guaranteed to be locally linear there.

In a similar way, we say that a two-variable function f is *locally linear near* (x_0, y_0) provided that the graph of f looks like a plane when viewed on a small scale near (x_0, y_0). There are, of course, functions that are not locally linear at some points (x_0, y_0). However, it turns out that if the first-order partial derivatives, f_x and f_y, are continuous near (x_0, y_0), then f is locally linear at (x_0, y_0) and the graph looks like a plane, which we call the tangent plane, when viewed on a small scale. Moreover, when a function is locally linear at a point, we will also say it is *differentiable* at that point.

If f is a function of the independent variables x and y and both f_x and f_y exist and are continuous in an open disk containing the point (x_0, y_0), then f is **differentiable** at (x_0, y_0).

So, whenever a function $z = f(x, y)$ is differentiable at a point (x_0, y_0), it follows that the function has a tangent plane at (x_0, y_0). Viewed up close, the tangent plane and the function are then virtually indistinguishable. In addition, as in Preview Activity 10.4, we find the following

10.4. LINEARIZATION: TANGENT PLANES AND DIFFERENTIALS

general formula for the tangent plane.

> If $f(x, y)$ has continuous first-order partial derivatives, then the equation of the plane tangent to the graph of f at the point $(x_0, y_0, f(x_0, y_0))$ is
>
> $$z = f(x_0, y_0) + f_x(x_0, y_0)(x - x_0) + f_y(x_0, y_0)(y - y_0). \tag{10.1}$$

Finally, one important note about the form of the equation for the tangent plane, $z = f(x_0, y_0) + f_x(x_0, y_0)(x - x_0) + f_y(x_0, y_0)(y - y_0)$. Say, for example, that we have the particular tangent plane $z = 7 - 2(x-3) + 4(y+1)$. Observe that we can immediately read from this form that $f_x(3, -1) = -2$ and $f_y(3, -1) = 4$; furthermore, $f_x(3, -1) = -2$ is the slope of the trace to both f and the tangent plane in the x-direction at $(-3, 1)$. In the same way, $f_y(3, -1) = 4$ is the slope of the trace of both f and the tangent plane in the y-direction at $(3, -1)$.

Activity 10.11.

Find the equation of the tangent plane to $f(x, y) = x^2 y$ at the point $(1, 2)$.

◁

Linearization

In single variable calculus, an important use of the tangent line is to approximate the value of a differentiable function. Near the point x_0, the tangent line to the graph of f at x_0 is close to the graph of f near x_0, as shown in Figure 10.30.

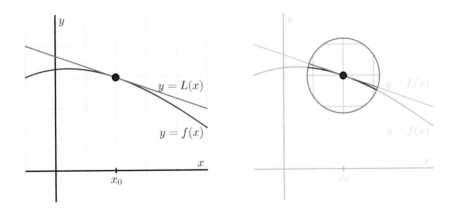

Figure 10.30: The linearization of the single-variable function $f(x)$.

In this single-variable setting, we let L denote the function whose graph is the tangent line, and thus

$$L(x) = f(x_0) + f'(x_0)(x - x_0)$$

Furthermore, observe that $f(x) \approx L(x)$ near x_0. We call L the *linearization* of f.

In the same way, the tangent plane to the graph of a differentiable function $z = f(x, y)$ at a point (x_0, y_0) provides a good approximation of $f(x, y)$ near (x_0, y_0). Here, we define the linearization, L, to be the two-variable function whose graph is the tangent plane, and thus

$$L(x, y) = f(x_0, y_0) + f_x(x_0, y_0)(x - x_0) + f_y(x_0, y_0)(y - y_0).$$

Finally, note that $f(x, y) \approx L(x, y)$ for points near (x_0, y_0). This is illustrated in Figure 10.31.

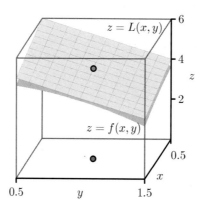

Figure 10.31: The linearization of $f(x, y)$.

Activity 10.12.

In what follows, we find the linearization of several different functions that are given in algebraic, tabular, or graphical form.

(a) Find the linearization $L(x, y)$ for the function g defined by

$$g(x, y) = \frac{x}{x^2 + y^2}$$

at the point $(1, 2)$. Then use the linearization to estimate the value of $g(0.8, 2.3)$.

(b) Table 10.3 provides a collection of values of the wind chill $w(v, T)$, in degrees Fahrenheit, as a function of wind speed, in miles per hour, and temperature, also in degrees Fahrenheit.

Use the data to first estimate the appropriate partial derivatives, and then find the linearization $L(v, T)$ at the point $(25, -10)$. Finally, use the linearization to estimate $w(25, -12)$, $w(23, -10)$, and $w(23, -12)$.

(c) Figure 10.32 gives a contour plot of a differentiable function f.

After estimating appropriate partial derivatives, determine the linearization $L(x, y)$ at the point $(2, 1)$, and use it to estimate $f(2.2, 1)$, $f(2, 0.8)$, and $f(2.2, 0.8)$.

◁

10.4. LINEARIZATION: TANGENT PLANES AND DIFFERENTIALS

$v\backslash T$	-30	-25	-20	-15	-10	-5	0	5	10	15	20
5	-46	-40	-34	-28	-22	-16	-11	-5	1	7	13
10	-53	-47	-41	-35	-28	-22	-16	-10	-4	3	9
15	-58	-51	-45	-39	-32	-26	-19	-13	-7	0	6
20	-61	-55	-48	-42	-35	-29	-22	-15	-9	-2	4
25	-64	-58	-51	-44	-37	-31	-24	-17	-11	-4	3
30	-67	-60	-53	-46	-39	-33	-26	-19	-12	-5	1
35	-69	-62	-55	-48	-41	-34	-27	-21	-14	-7	0
40	-71	-64	-57	-50	-43	-36	-29	-22	-15	-8	-1

Table 10.3: Wind chill as a function of wind speed and temperature.

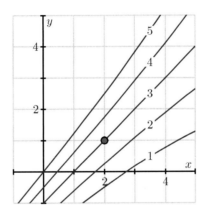

Figure 10.32: A contour plot of $f(x, y)$.

Differentials

As we have seen, the linearization $L(x, y)$ enables us to estimate the value of $f(x, y)$ for points (x, y) near the base point (x_0, y_0). Sometimes, however, we are more interested in the *change* in f as we move from the base point (x_0, y_0) to another point (x, y).

Figure 10.33 illustrates this situation. Suppose we are at the point (x_0, y_0), and we know the value $f(x_0, y_0)$ of f at (x_0, y_0). If we consider the displacement $\langle \Delta x, \Delta y \rangle$ to a new point $(x, y) = (x_0 + \Delta x, y_0 + \Delta y)$, we would like to know how much the function has changed. We denote this change by Δf, where

$$\Delta f = f(x, y) - f(x_0, y_0).$$

A simple way to estimate the change Δf is to approximate it by df, which represents the change

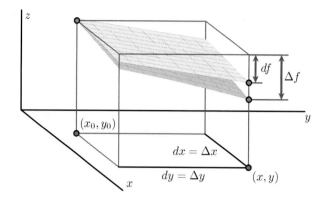

Figure 10.33: The differential df measures the approximate change in $f(x, y)$.

in the linearization $L(x, y)$ as we move from (x_0, y_0) to (x, y). This gives

$$\begin{aligned}\Delta f \approx df &= L(x, y) - f(x_0, y_0) \\ &= [f(x_0, y_0) + f_x(x_0, y_0)(x - x_0) + f_y(x_0, y_0)(y - y_0)] - f(x_0, y_0) \\ &= f_x(x_0, y_0)\Delta x + f_y(x_0, y_0)\Delta y.\end{aligned}$$

For consistency, we will denote the change in the independent variables as $dx = \Delta x$ and $dy = \Delta y$, and thus

$$\Delta f \approx df = f_x(x_0, y_0)\, dx + f_y(x_0, y_0)\, dy. \tag{10.2}$$

Expressed equivalently in Leibniz notation, we have

$$df = \frac{\partial f}{\partial x}\, dx + \frac{\partial f}{\partial y}\, dy. \tag{10.3}$$

We call the quantities dx, dy, and df *differentials*, and we think of them as measuring small changes in the quantities x, y, and f. Equations (10.2) and (10.3) express the relationship between these changes. Equation (10.3) resembles an important idea from single-variable calculus: when y depends on x, it follows in the notation of differentials that

$$dy = y'\, dx = \frac{dy}{dx}\, dx.$$

We will illustrate the use of differentials with an example. Suppose we have a machine that manufactures rectangles of width $x = 20$ cm and height $y = 10$ cm. However, the machine isn't perfect, and therefore the width could be off by $dx = \Delta x = 0.2$ cm and the height could be off by $dy = \Delta y = 0.4$ cm.

The area of the rectangle is

$$A(x, y) = xy,$$

so that the area of a perfectly manufactured rectangle is $A(20, 10) = 200$ square centimeters. Since the machine isn't perfect, we would like to know how much the area of a given manufactured rectangle could differ from the perfect rectangle. We will estimate the uncertainty in the area using (10.2), and find that

$$\Delta A \approx dA = A_x(20, 10)\, dx + A_y(20, 10)\, dy.$$

Since $A_x = y$ and $A_y = x$, we have

$$\Delta A \approx dA = 10\, dx + 20\, dy = 10 \cdot 0.2 + 20 \cdot 0.4 = 10.$$

That is, we estimate that the area in our rectangles could be off by as much as 10 square centimeters.

Activity 10.13.

The questions in this activity explore the differential in several different contexts.

(a) Suppose that the elevation of a landscape is given by the function h, where we additionally know that $h(3,1) = 4.35$, $h_x(3,1) = 0.27$, and $h_y(3,1) = -0.19$. Assume that x and y are measured in miles in the easterly and northerly directions, respectively, from some base point $(0,0)$.

Your GPS device says that you are currently at the point $(3,1)$. However, you know that the coordinates are only accurate to within 0.2 units; that is, $dx = \Delta x = 0.2$ and $dy = \Delta y = 0.2$. Estimate the uncertainty in your elevation using differentials.

(b) The pressure, volume, and temperature of an ideal gas are related by the equation

$$P = P(T, V) = 8.31 T/V,$$

where P is measured in kilopascals, V in liters, and T in kelvin. Find the pressure when the volume is 12 liters and the temperature is 310 K. Use differentials to estimate the change in the pressure when the volume increases to 12.3 liters and the temperature decreases to 305 K.

(c) Refer to Table 10.3, the table of values of the wind chill $w(v, T)$, in degrees Fahrenheit, as a function of temperature, also in degrees Fahrenheit, and wind speed, in miles per hour.

Suppose your anemometer says the wind is blowing at 25 miles per hour and your thermometer shows a reading of $-15°$ degrees. However, you know your thermometer is only accurate to within $2°$ degrees and your anemometer is only accurate to within 3 miles per hour. What is the wind chill based on your measurements? Estimate the uncertainty in your measurement of the wind chill.

◁

Summary

- A function f of two independent variables is locally linear at a point (x_0, y_0) if the graph of f looks like a plane as we zoom in on the graph around the point (x_0, y_0). In this case, the equation of the tangent plane is given by

$$z = f(x_0, y_0) + f_x(x_0, y_0)(x - x_0) + f_y(x_0, y_0)(y - y_0).$$

- The tangent plane $L(x, y) = f(x_0, y_0) + f_x(x_0, y_0)(x - x_0) + f_y(x_0, y_0)(y - y_0)$, when considered as a function, is called the linearization of a differentiable function f at (x_0, y_0) and may be used to estimate values of $f(x, y)$; that is, $f(x, y) \approx L(x, y)$ for points (x, y) near (x_0, y_0).

- A function f of two independent variables is differentiable at (x_0, y_0) provided that both f_x and f_y exist and are continuous in an open disk containing the point (x_0, y_0).

- The differential df of a function $f = f(x, y)$ is related to the differentials dx and dy by

$$df = f_x(x_0, y_0)dx + f_y(x_0, y_0)dy.$$

We can use this relationship to approximate small changes in f that result from small changes in x and y.

Exercises

1. Let f be the function defined by $f(x, y) = x^{1/3}y^{1/3}$, whose graph is shown in Figure 10.34.

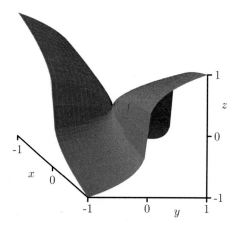

Figure 10.34: The surface for $f(x, y) = x^{1/3}y^{1/3}$.

(a) Determine
$$\lim_{h \to 0} \frac{f(0 + h, 0) - f(0, 0)}{h}.$$
What does this limit tell us about $f_x(0, 0)$?

(b) Note that $f(x, y) = f(y, x)$, and this symmetry implies that $f_x(0, 0) = f_y(0, 0)$. So both partial derivatives of f exist at $(0, 0)$. A picture of the surface defined by f near $(0, 0)$ is shown in Figure 10.34. Based on this picture, do you think f is locally linear at $(0, 0)$? Why?

(c) Show that the curve where $x = y$ on the surface defined by f is not differentiable at 0. What does this tell us about the local linearity of f at $(0, 0)$?

10.4. LINEARIZATION: TANGENT PLANES AND DIFFERENTIALS

(d) Is the function f defined by $f(x,y) = \frac{x^2}{y^2+1}$ locally linear at $(0,0)$? Why or why not?

2. Let g be a function that is differentiable at $(-2, 5)$ and suppose that its tangent plane at this point is given by $z = -7 + 4(x+2) - 3(y-5)$.

   (a) Determine the values of $g(-2,5)$, $g_x(-2,5)$, and $g_y(-2,5)$. Write one sentence to explain your thinking.

   (b) Estimate the value of $g(-1.8, 4.7)$. Clearly show your work and thinking.

   (c) Given changes of $dx = -0.34$ and $dy = 0.21$, estimate the corresponding change in g that is given by its differential, dg.

   (d) Suppose that another function h is also differentiable at $(-2, 5)$, but that its tangent plane at $(-2, 5)$ is given by $3x + 2y - 4z = 9$. Determine the values of $h(-2,5)$, $h_x(-2,5)$, and $h_y(-2,5)$, and then estimate the value of $h(-1.8, 4.7)$. Clearly show your work and thinking.

3. In the following questions, we determine and apply the linearization for several different functions.

   (a) Find the linearization $L(x, y)$ for the function f defined by $f(x,y) = \cos(x)(2e^{2y} + e^{-2y})$ at the point $(x_0, y_0) = (0, 0)$. Hence use the linearization to estimate the value of $f(0.1, 0.2)$. Compare your estimate to the actual value of $f(0.1, 0.2)$.

   (b) The Heat Index, I, (measured in apparent degrees F) is a function of the actual temperature T outside (in degrees F) and the relative humidity H (measured as a percentage). A portion of the table which gives values for this function, $I = I(T, H)$, is provided below:

$T \downarrow \backslash H \rightarrow$	70	75	80	85
90	106	109	112	115
92	112	115	119	123
94	118	122	127	132
96	125	130	135	141

   Suppose you are given that $I_T(94, 75) = 3.75$ and $I_H(94, 75) = 0.9$. Use this given information and one other value from the table to estimate the value of $I(93.1, 77)$ using the linearization at $(94, 75)$. Using proper terminology and notation, explain your work and thinking.

   (c) Just as we can find a local linearization for a differentiable function of two variables, we can do so for functions of three or more variables. By extending the concept of the local linearization from two to three variables, find the linearization of the function $h(x, y, z) = e^{2x}(y + z^2)$ at the point $(x_0, y_0, z_0) = (0, 1, -2)$. Then, use the linearization to estimate the value of $h(-0.1, 0.9, -1.8)$.

4. In the following questions, we investigate two different applied settings using the differential.

(a) Let f represent the vertical displacement in centimeters from the rest position of a string (like a guitar string) as a function of the distance x in centimeters from the fixed left end of the string and y the time in seconds after the string has been plucked.[2] A simple model for f could be
$$f(x, y) = \cos(x)\sin(2y).$$
Use the differential to approximate how much more this vibrating string is vertically displaced from its position at $(a, b) = \left(\frac{\pi}{4}, \frac{\pi}{3}\right)$ if we decrease a by 0.01 cm and increase the time by 0.1 seconds. Compare to the value of f at the point $\left(\frac{\pi}{4} - 0.01, \frac{\pi}{3} + 0.1\right)$.

(b) Resistors used in electrical circuits have colored bands painted on them to indicate the amount of resistance and the possible error in the resistance. When three resistors, whose resistances are R_1, R_2, and R_3, are connected in parallel, the total resistance R is given by
$$\frac{1}{R} = \frac{1}{R_1} + \frac{1}{R_2} + \frac{1}{R_3}.$$
Suppose that the resistances are $R_1 = 25\Omega$, $R_2 = 40\Omega$, and $R_3 = 50\Omega$. Find the total resistance R.

If you know each of R_1, R_2, and R_3 with a possible error of 0.5%, estimate the maximum error in your calculation of R.

---

[2] An interesting video of this can be seen at https://www.youtube.com/watch?v=TKF6nFzpHBUA.

10.5 The Chain Rule

Motivating Questions

In this section, we strive to understand the ideas generated by the following important questions:

- What is the Chain Rule and how do we use it to find a derivative?
- How can we use a tree diagram to guide us in applying the Chain Rule?

Introduction

In single-variable calculus, we encountered situations in which some quantity z depends on y and, in turn, y depends on x. A change in x produces a change in y, which consequently produces a change in z. Using the language of differentials that we saw in the previous section, these changes are naturally related by

$$dz = \frac{dz}{dy} dy \quad \text{and} \quad dy = \frac{dy}{dx} dx.$$

In terms of instantaneous rates of change, we then have

$$dz = \frac{dz}{dy}\frac{dy}{dx} dx = \frac{dz}{dx} dx$$

and thus

$$\frac{dz}{dx} = \frac{dz}{dy}\frac{dy}{dx}.$$

This most recent equation we call the *Chain Rule*.

In the case of a function f of two variables where $z = f(x, y)$, it might be that both x and y depend on another variable t. A change in t then produces changes in both x and y, which then cause z to change. In this section we will see how to find the change in z that is caused by a change in t, leading us to multivariable versions of the Chain Rule involving both regular and partial derivatives.

Preview Activity 10.5. Suppose you are driving around in the x-y plane in such a way that your position $\mathbf{r}(t)$ at time t is given by function

$$\mathbf{r}(t) = \langle x(t), y(t) \rangle = \langle 2 - t^2, t^3 + 1 \rangle.$$

The path taken is shown on the left of Figure 10.35.

Suppose, furthermore, that the temperature at a point in the plane is given by

$$T(x, y) = 10 - \frac{1}{2}x^2 - \frac{1}{5}y^2,$$

and note that the surface generated by T is shown on the right of Figure 10.35. Therefore, as time passes, your position $(x(t), y(t))$ changes, and, as your position changes, the temperature $T(x, y)$ also changes.

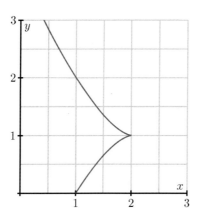
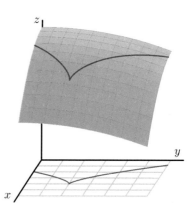

Figure 10.35: At left, your position in the plane; at right, the corresponding temperature.

(a) The position function **r** provides a parameterization $x = x(t)$ and $y = y(t)$ of the position at time t. By substituting $x(t)$ for x and $y(t)$ for y in the formula for T, we can write $T = T(x(t), y(t))$ as a function of t. Make these substitutions to write T as a function of t and then use the Chain Rule from single variable calculus to find $\frac{dT}{dt}$. (Do not do any algebra to simplify the derivative, either before taking the derivative, nor after.)

(b) Now we want to understand how the result from part (a) can be obtained from T as a multivariable function. Recall from the previous section that small changes in x and y produce a change in T that is approximated by

$$\Delta T \approx T_x \Delta x + T_y \Delta y.$$

The Chain Rule tells us about the instantaneous rate of change of T, and this can be found as

$$\lim_{\Delta t \to 0} \frac{\Delta T}{\Delta t} = \lim_{\Delta t \to 0} \frac{T_x \Delta x + T_y \Delta y}{\Delta t}. \tag{10.4}$$

Use equation (10.4) to explain why the instantaneous rate of change of T that results from a change in t is

$$\frac{dT}{dt} = \frac{\partial T}{\partial x}\frac{dx}{dt} + \frac{\partial T}{\partial y}\frac{dy}{dt}. \tag{10.5}$$

(c) Using the original formulas for T, x, and y in the problem statement, calculate all of the derivatives in Equation (10.5) (with T_x and T_y in terms of x and y, and x' and y' in terms of t), and hence write the right-hand side of Equation (10.5) in terms of x, y, and t.

(d) Compare the results of parts (a) and (c). Write a couple of sentences that identify specifically how each term in (c) relates to a corresponding terms in (a). This connection between parts (a) and (c) provides a multivariable version of the Chain Rule.

10.5. THE CHAIN RULE

The Chain Rule

As Preview Activity 10.3 suggests, the following version of the Chain Rule holds in general.

> Let $z = f(x, y)$, where f is a differentiable function of the independent variables x and y, and let x and y each be differentiable functions of an independent variable t. Then
> $$\frac{dz}{dt} = \frac{\partial z}{\partial x}\frac{dx}{dt} + \frac{\partial z}{\partial y}\frac{dy}{dt}. \tag{10.6}$$

It is important to note the differences among the derivatives in (10.6). Since z is a function of the two variables x and y, the derivatives in the Chain Rule for z with respect to x and y are partial derivatives. However, since $x = x(t)$ and $y = y(t)$ are functions of the single variable t, their derivatives are the standard derivatives of functions of one variable. When we compose z with $x(t)$ and $y(t)$, we then have z as a function of the single variable t, making the derivative of z with respect to t a standard derivative from single variable calculus as well.

To understand why this Chain Rule works in general, suppose that some quantity z depends on x and y so that

$$dz = \frac{\partial z}{\partial x}dx + \frac{\partial z}{\partial y}dy. \tag{10.7}$$

Next, suppose that x and y each depend on another quantity t, so that

$$dx = \frac{dx}{dt}dt \quad \text{and} \quad dy = \frac{dy}{dt}dt. \tag{10.8}$$

Combining Equations (10.7) and (10.8), we find that

$$dz = \frac{\partial z}{\partial x}\frac{dx}{dt}dt + \frac{\partial z}{\partial y}\frac{dy}{dt}dt = \frac{dz}{dt}dt,$$

which is the Chain Rule in this particular context, as expressed in Equation 10.6.

Activity 10.14.

In the following questions, we apply the recently-developed Chain Rule in several different contexts.

(a) Suppose that we have a function z defined by $z(x, y) = x^2 + xy^3$. In addition, suppose that x and y are restricted to points that move around the plane by following a circle of radius 2 centered at the origin that is parameterized by

$$x(t) = 2\cos(t), \quad \text{and} \quad y(t) = 2\sin(t).$$

Use the Chain Rule to find the resulting instantaneous rate of change $\frac{dz}{dt}$.

(b) Suppose that the temperature on a metal plate is given by the function T with
$$T(x,y) = 100 - (x^2 + 4y^2),$$
where the temperature is measured in degrees Fahrenheit and x and y are each measured in feet.

   i. Find T_x and T_y. What are the units on these partial derivatives?
   ii. Suppose an ant is walking along the x-axis at the rate of 2 feet per minute toward the origin. When the ant is at the point $(2,0)$, what is the instantaneous rate of change in the temperature dT/dt that the ant experiences. Include units on your response.
   iii. Suppose instead that the ant walks along an ellipse with $x = 6\cos(t)$ and $y = 3\sin(t)$, where t is measured in minutes. Find $\frac{dT}{dt}$ at $t = \pi/6$, $t = \pi/4$, and $t = \pi/3$. What does this seem to tell you about the path along which the ant is walking?

(c) Suppose that you are walking along a surface whose elevation is given by a function f. Furthermore, suppose that if you consider how your location corresponds to points in the x-y plane, you know that when you pass the point $(2,1)$, your velocity vector is $\mathbf{v} = \langle -1, 2 \rangle$. If some contours of f are as shown in Figure 10.36, estimate the rate of change df/dt when you pass through $(2,1)$.

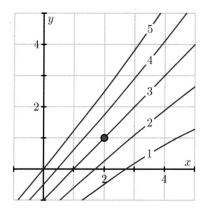

Figure 10.36: Some contours of f.

Tree Diagrams

Up to this point, we have applied the Chain Rule to situations where we have a function z of variables x and y, with both x and y depending on another single quantity t. We may apply the Chain Rule, however, when x and y each depend on more than one quantity, or when z is a function of more than two variables. It can be challenging to keep track of all the dependencies

10.5. THE CHAIN RULE

among the variables, and thus a tree diagram can be a useful tool to organize our work. For example, suppose that z depends on x and y, and x and y both depend on t. We may represent these relationships using the tree diagram shown at left Figure 10.37. We place the dependent variable at the top of the tree and connect it to the variables on which it depends one level below. We then connect each of those variables to the variable on which each depends.

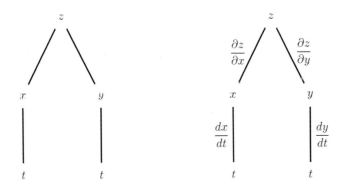

Figure 10.37: A tree diagram illustrating dependencies.

To represent the Chain Rule, we label every edge of the diagram with the appropriate derivative or partial derivative, as seen at right in Figure 10.37. To calculate an overall derivative according to the Chain Rule, we construct the product of the derivatives along all paths connecting the variables and then add all of these products. For example, the diagram at right in 10.37 illustrates the Chain Rule

$$\frac{dz}{dt} = \frac{\partial z}{\partial x}\frac{dx}{dt} + \frac{\partial z}{\partial y}\frac{dy}{dt}.$$

Activity 10.15.

(a) Figure 10.38 shows the tree diagram we construct when (a) z depends on w, x, and y, (b) w, x, and y each depend on u and v, and (c) u and v depend on t.

  i. Label the edges with the appropriate derivatives.
  ii. Use the Chain Rule to write $\dfrac{dz}{dt}$.

(b) Suppose that $z = x^2 - 2xy^2$ and that

$$x = r\cos(\theta)$$
$$y = r\sin(\theta).$$

  i. Construct a tree diagram representing the dependencies of z on x and y and x and y on r and θ.
  ii. Use the tree diagram to find $\frac{\partial z}{\partial r}$.
  iii. Now suppose that $r = 3$ and $\theta = \pi/6$. Find the values of x and y that correspond to these given values of r and θ, and then use the Chain Rule to find the value of the partial derivative $\frac{\partial z}{\partial \theta}\big|_{(3,\frac{\pi}{6})}$.

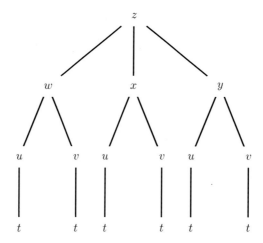

Figure 10.38: Three levels of dependencies

Summary

- The Chain Rule is a tool for differentiating a composite for functions. In its simplest form, it says that if $f(x, y)$ is a function of two variables and $x(t)$ and $y(t)$ depend on t, then

$$\frac{df}{dt} = \frac{\partial f}{\partial x}\frac{dx}{dt} + \frac{\partial f}{\partial y}\frac{dy}{dt}.$$

- A tree diagram can be used to represent the dependence of variables on other variables. By following the links in the tree diagram, we can form chains of partial derivatives or derivatives that can be combined to give a desired partial derivative.

Exercises

1. Find the indicated derivative. In each case, state the version of the Chain Rule that you are using.

    (a) $\frac{df}{dt}$, if $f(x, y) = 2x^2 y$, $x = \cos(t)$, and $y = \ln(t)$.

    (b) $\frac{\partial f}{\partial w}$, if $f(x, y) = 2x^2 y$, $x = w + z^2$, and $y = \frac{2z+1}{w}$.

    (c) $\frac{\partial f}{\partial v}$, if $f(x, y, z) = 2x^2 y + z^3$, $x = u - v + 2w$, $y = w2^v - u^3$, and $z = u^2 - v$.

2. Let $z = u^2 - v^2$ and suppose that

$$u = e^x \cos(y)$$
$$v = e^x \sin(y)$$

(a) Find the values of u and v that correspond to $x = 0$ and $y = 2\pi/3$.

(b) Use the Chain Rule to find the general partial derivatives

$$\frac{\partial z}{\partial x} \quad \text{and} \quad \frac{\partial z}{\partial y}$$

and then determine both $\frac{\partial z}{\partial x}\big|_{(0,\frac{2\pi}{3})}$ and $\frac{\partial z}{\partial y}\big|_{(0,\frac{2\pi}{3})}$.

3. Suppose that $T = x^2 + y^2 - 2z$ where

$$x = \rho \sin(\phi) \cos(\theta)$$
$$y = \rho \sin(\phi) \sin(\theta)$$
$$z = \rho \cos(\phi)$$

(a) Construct a tree diagram representing the dependencies among the variables.

(b) Apply the chain rule to find the partial derivatives

$$\frac{\partial T}{\partial \rho}, \frac{\partial T}{\partial \phi}, \text{ and } \frac{\partial T}{\partial \theta}.$$

4. Suppose that the temperature on a metal plate is given by the function T with

$$T(x, y) = 100 - (x^2 + 4y^2),$$

where the temperature is measured in degrees Fahrenheit and x and y are each measured in feet. Now suppose that an ant is walking on the metal plate in such a way that it walks in a straight line from the point $(1, 4)$ to the point $(5, 6)$.

(a) Find parametric equations $(x(t), y(t))$ for the ant's coordinates as it walks the line from $(1, 4)$ to $(5, 6)$.

(b) What can you say about $\frac{dx}{dt}$ and $\frac{dy}{dt}$ for every value of t?

(c) Determine the instantaneous rate of change in temperature with respect to t that the ant is experiencing at the moment it is halfway from $(1, 4)$ to $(5, 6)$, using your parametric equations for x and y. Include units on your answer.

5. There are several proposed formulas to approximate the surface area of the human body. One model[3] uses the formula

$$A(h, w) = 0.0072 h^{0.725} w^{0.425},$$

where A is the surface area in square meters, h is the height in centimeters, and w is the weight in kilograms.

---

[3]DuBois D, DuBois DF. A formula to estimate the approximate surface area if height and weight be known. *Arch Int Med* 1916;17:863-71.

Since a person's height h and weight w change over time, h and w are functions of time t. Let us think about what is happening to a child whose height is 60 centimeters and weight is 9 kilograms. Suppose, furthermore, that h is increasing at an instantaneous rate of 20 centimeters per year and w is increasing at an instantaneous rate of 5 kg per year.

Determine the instantaneous rate at which the child's surface area is changing at this point in time.

6. Let $z = f(x,y) = 50 - (x+1)^2 - (y+3)^2$ and $z = h(x,y) = 24 - 2x - 6y$.

   Suppose a person is walking on the surface $z = f(x,y)$ in such a way that she walks the curve which is the intersection of f and h.

   (a) Show that $x(t) = 4\cos(t)$ and $y(t) = 4\sin(t)$ is a parameterization of the "shadow" in the x-y plane of the curve that is the intersection of the graphs of f and h.

   (b) Use the parameterization from part (a) to find the instantaneous rate at which her height is changing with respect to time at the instant $t = 2\pi/3$.

7. The voltage V (in volts) across a circuit is given by Ohm's Law: $V = IR$, where I is the current (in amps) in the circuit and R is the resistance (in ohms). Suppose we connect two resistors with resistances R_1 and R_2 in parallel as shown in Figure 10.39. The total resistance R in the circuit is then given by

$$\frac{1}{R} = \frac{1}{R_1} + \frac{1}{R_2}.$$

   (a) Assume that the current, I, and the resistances, R_1 and R_2, are changing over time, t. Use the Chain Rule to write a formula for $\frac{dV}{dt}$.

   (b) Suppose that, at some particular point in time, we measure the current to be 3 amps and that the current is increasing at $\frac{1}{10}$ amps per second, while resistance R_1 is 2 ohms and decreasing at the rate of 0.2 ohms per second and R_2 is 1 ohm and increasing at the rate of 0.5 ohms per second. At what rate is the voltage changing at this point in time?

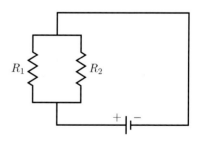

Figure 10.39: Resistors in parallel.

10.6 Directional Derivatives and the Gradient

Motivating Questions

In this section, we strive to understand the ideas generated by the following important questions:

- The partial derivatives of a function f tell us the rate of change of f in the direction of the coordinate axes. How can we measure the rate of change of f in other directions?

- What is the gradient of a function and what does it tell us?

Introduction

The partial derivatives of a function tell us the instantaneous rate at which the function changes as we hold all but one independent variable constant and allow the remaining independent variable to change. It is natural to wonder how we can measure the rate at which a function changes in directions other than parallel to a coordinate axes. In what follows, we investigate this question, and see how the rate of change in any given direction is connected to the rates of change given by the standard partial derivatives.

Preview Activity 10.6. Let's consider the function f defined by

$$f(x, y) = 30 - x^2 - \frac{1}{2}y^2,$$

and suppose that f measures the temperature, in degrees Celsius, at a given point in the plane, where x and y are measured in feet. Assume that the positive x-axis points due east, while the positive y-axis points due north. A contour plot of f is shown in Figure 10.40

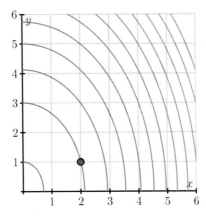

Figure 10.40: A contour plot of $f(x, y) = 30 - x^2 - \frac{1}{2}y^2$.

(a) Suppose that a person is walking due east, and thus parallel to the x-axis. At what instantaneous rate is the temperature changing at the moment she passes the point $(2, 1)$? What are the units on this rate of change?

(b) Next, determine the instantaneous rate of change of temperature at the point $(2, 1)$ if the person is instead walking due north. Again, include units on your result.

(c) Now, rather than walking due east or due north, let's suppose that the person is walking with velocity given by the vector $\mathbf{v} = \langle 3, 4 \rangle$, where time is measured in seconds. Note that the person's speed is thus $|\mathbf{v}| = 5$ feet per second.

Find parametric equations for the person's path; that is, parameterize the line through $(2, 1)$ using the direction vector $\mathbf{v} = \langle 3, 4 \rangle$. Let $x(t)$ denote the x-coordinate of the line, and $y(t)$ its y-coordinate.

(d) With the parameterization in (c), we can now view the temperature f as not only a function of x and y, but also of time, t. Hence, use the chain rule to determine the value of $\frac{df}{dt}\big|_{t=0}$. What are the units on your answer? What is the practical meaning of this result?

Directional Derivatives

Given a function $z = f(x, y)$, the partial derivative $f_x(x_0, y_0)$ measures the instantaneous rate of change of f as only the x variable changes; likewise, $f_y(x_0, y_0)$ measures the rate of change of f at (x_0, y_0) as only y changes. Note particularly that $f_x(x_0, y_0)$ is measured in "units of f per unit of change in x," and that the units on $f_y(x_0, y_0)$ are similar.

In Preview Activity 10.6, we saw how we could measure the rate of change of f in a situation where both x and y were changing; in that activity, however, we found that this rate of change was measured in "units of f per unit of *time*." In a given unit of time, we may move more than one unit of distance. In fact, in Preview Activity 10.6, in each unit increase in time we move a distance of $|\mathbf{v}| = 5$ feet. To generalize the notion of partial derivatives to any direction of our choice, we instead want to have a rate of change whose units are "units of f per unit of distance in the given direction."

In this light, in order to formally define the derivative in a particular direction of motion, we want to represent the change in f for a given *unit* change in the direction of motion. We can represent this unit change in direction with a unit vector, say $\mathbf{u} = \langle u_1, u_2 \rangle$. If we move a distance h in the direction of $\mathbf{u}$ from a fixed point (x_0, y_0), we then arrive at the new point $(x_0 + u_1 h, y_0 + u_2 h)$. It now follows that the slope of the secant line to the curve on the surface through (x_0, y_0) in the direction of $\mathbf{u}$ through the points (x_0, y_0) and $(x_0 + u_1 h, y_0 + u_2 h)$ is

$$m_{\text{sec}} = \frac{f(x_0 + u_1 h, y_0 + u_2 h) - f(x_0, y_0)}{h}. \tag{10.9}$$

10.6. DIRECTIONAL DERIVATIVES AND THE GRADIENT

To get the instantaneous rate of change of f in the direction $\mathbf{u} = \langle u_1, u_2 \rangle$, we must take the limit of the quantity in Equation (10.9) as $h \to 0$. Doing so results in the formal definition of the directional derivative.

Definition 10.4. Let $f = f(x, y)$ be given. The **derivative of f at the point (x, y) in the direction of the unit vector** $\mathbf{u} = \langle u_1, u_2 \rangle$ is denoted $D_{\mathbf{u}} f(x, y)$ and is given by

$$D_{\mathbf{u}} f(x, y) = \lim_{h \to 0} \frac{f(x + u_1 h, y + u_2 h) - f(x, y)}{h} \tag{10.10}$$

for those values of x and y for which the limit exists.

The quantity $D_{\mathbf{u}} f(x, y)$ is called a *directional derivative*. When we evaluate the directional derivative $D_{\mathbf{u}} f(x, y)$ at a point (x_0, y_0), the result $D_{\mathbf{u}} f(x_0, y_0)$ tells us the instantaneous rate at which f changes at (x_0, y_0) per unit increase in the direction of the vector $\mathbf{u}$. In addition, the quantity $D_{\mathbf{u}} f(x_0, y_0)$ tells us the slope of the line tangent to the surface in the direction of $\mathbf{u}$ at the point $(x_0, y_0, f(x_0, y_0))$.

Computing the Directional Derivative

In a similar way to how we developed shortcut rules for standard derivatives in single variable calculus, and for partial derivatives in multivariable calculus, we can also find a way to evaluate directional derivatives without resorting to the limit definition found in Equation (10.10). We do so using a very similar approach to our work in Preview Activity 10.6.

Suppose we consider the situation where we are interested in the instantaneous rate of change of f at a point (x_0, y_0) in the direction $\mathbf{u} = \langle u_1, u_2 \rangle$, where $\mathbf{u}$ is a unit vector. The variables x and y are therefore changing according to the parameterization

$$x = x_0 + u_1 t \quad \text{and} \quad y = y_0 + u_2 t.$$

Observe that $\frac{dx}{dt} = u_1$ and $\frac{dy}{dt} = u_2$ for all values of t. Since $\mathbf{u}$ is a unit vector, it follows that a point moving along this line moves one unit of distance per one unit of time; that is, each single unit of time corresponds to movement of a single unit of distance in that direction. This observation allows us to use the Chain Rule to calculate the directional derivative, which measures the instantaneous rate of change of f with respect to change in the direction $\mathbf{u}$.

In particular, by the Chain Rule, it follows that

$$D_{\mathbf{u}} f(x_0, y_0) = f_x(x_0, y_0) \frac{dx}{dt}\bigg|_{(x_0, y_0)} + f_y(x_0, y_0) \frac{dy}{dt}\bigg|_{(x_0, y_0)} = f_x(x_0, y_0) u_1 + f_y(x_0, y_0) u_2.$$

This now allows us to compute the directional derivative at an arbitrary point according to the following formula.

> Given a differentiable function $f = f(x,y)$ and a unit vector $\mathbf{u} = \langle u_1, u_2 \rangle$, we may compute $D_{\mathbf{u}}f(x,y)$ by
> $$D_{\mathbf{u}}f(x,y) = f_x(x,y)u_1 + f_y(x,y)u_2. \tag{10.11}$$

Note well: To use Equation (10.11), we must have a *unit* vector $\mathbf{u} = \langle u_1, u_2 \rangle$ in the direction of motion. In the event that we have a direction prescribed by a non-unit vector, we must first scale the vector to have length 1.

Activity 10.16.

Let $f(x,y) = 3xy - x^2y^3$.

(a) Determine $f_x(x,y)$ and $f_y(x,y)$.

(b) Use Equation (10.11) to determine $D_{\mathbf{i}}f(x,y)$ and $D_{\mathbf{j}}f(x,y)$. What familiar function is $D_{\mathbf{i}}f$? What familiar function is $D_{\mathbf{j}}f$?

(c) Use Equation (10.11) to find the derivative of f in the direction of the vector $\mathbf{v} = \langle 2, 3 \rangle$ at the point $(1, -1)$. Remember that a unit direction vector is needed.

◁

The Gradient

Via the Chain Rule, we have seen that for a given function $f = f(x,y)$, its instantaneous rate of change in the direction of a unit vector $\mathbf{u} = \langle u_1, u_2 \rangle$ is given by

$$D_{\mathbf{u}}f(x_0, y_0) = f_x(x_0, y_0)u_1 + f_y(x_0, y_0)u_2. \tag{10.12}$$

Recalling that the dot product of two vectors $\mathbf{v} = \langle v_1, v_2 \rangle$ and $\mathbf{u} = \langle u_1, u_2 \rangle$ is computed by

$$\mathbf{v} \cdot \mathbf{u} = v_1 u_1 + v_2 u_2,$$

we see that we may recast Equation (10.12) in a way that has geometric meaning. In particular, we see that $D_{\mathbf{u}}f(x_0, y_0)$ is the dot product of the vector $\langle f_x(x_0, y_0), f_y(x_0, y_0) \rangle$ and the vector $\mathbf{u}$.

We call this vector formed by the partial derivatives of f the *gradient* of f and denote it

$$\nabla f(x_0, y_0) = \langle f_x(x_0, y_0), f_y(x_0, y_0) \rangle.$$

We read ∇f as "the gradient of f," "grad f" or "del f".[4] Notice that ∇f varies from point to point. In the following activity, we investigate some of what the gradient tells us about the behavior of a function f.

---

[4]The symbol ∇ is called *nabla*, which comes from a Greek word for a certain type of harp that has a similar shape.

Activity 10.17.

Let's consider the function f defined by $f(x,y) = x^2 - y^2$. Some contours for this function are shown in Figure 10.41.

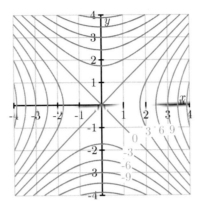

Figure 10.41: Contours of $f(x,y) = x^2 - y^2$.

(a) Find the gradient $\nabla f(x,y)$.

(b) For each of the following points (x_0, y_0), evaluate the gradient $\nabla f(x_0, y_0)$ and sketch the gradient vector with its tail at (x_0, y_0). Some of the vectors are too long to fit onto the plot, but we'd like to draw them to scale; to do so, scale each vector by a factor of $1/4$.

- $(x_0, y_0) = (2, 0)$
- $(x_0, y_0) = (0, 2)$
- $(x_0, y_0) = (2, 2)$
- $(x_0, y_0) = (2, 1)$
- $(x_0, y_0) = (-3, 2)$
- $(x_0, y_0) = (-2, -4)$
- $(x_0, y_0) = (0, 0)$

(c) What do you notice about the relationship between the gradient at (x_0, y_0) and the contour line passing through that point?

(d) Does f increase or decrease in the direction of $\nabla f(x_0, y_0)$? Provide a justification for your response.

◁

As a vector, $\nabla f(x_0, y_0)$ defines a direction and a length. As we will soon see, both of these convey important information about the behavior of f near (x_0, y_0).

The Direction of the Gradient

Remember that the dot product also conveys information about the angle between the two vectors. If θ is the angle between $\nabla f(x_0, y_0)$ and $\mathbf{u}$ (where $\mathbf{u}$ is a unit vector), then we also have that

$$D_{\mathbf{u}}f(x_0, y_0) = \nabla f(x_0, y_0) \cdot \mathbf{u} = |\nabla f(x_0, y_0)||\mathbf{u}|\cos(\theta).$$

In particular, when θ is a right angle, as shown on the left of Figure 10.42, then $D_{\mathbf{u}}f(x_0, y_0) = 0$, because $\cos(\theta) = 0$. Since the value of the directional derivative is 0, this means that f is unchanging in this direction, and hence $\mathbf{u}$ must be tangent to the contour of f that passes through (x_0, y_0). In other words, $\nabla f(x_0, y_0)$ is orthogonal to the contour through (x_0, y_0). This shows that the gradient vector at a given point is always perpendicular to the contour passing through the point, confirming that what we saw in part (c) of Activity 10.17 holds in general.

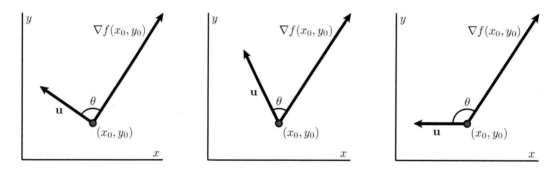

Figure 10.42: The sign of $D_{\mathbf{u}}f(x_0, y_0)$ is determined by θ.

Moreover, when θ is an acute angle, it follows that $\cos(\theta) > 0$ so since

$$D_{\mathbf{u}}f(x_0, y_0) = |\nabla f(x_0, y_0)||\mathbf{u}|\cos(\theta),$$

and therefore $D_{\mathbf{u}}f(x_0, y_0) > 0$, as shown in the middle image in Figure 10.42. This means that f is increasing in any direction where θ is acute. In a similar way, when θ is an obtuse angle, then $\cos(\theta) < 0$ so $D_{\mathbf{u}}f(x_0, y_0) < 0$, as seen on the right in Figure 10.42. This means that f is decreasing in any direction for which θ is obtuse.

Finally, as we can see in the following activity, we may also use the gradient to determine the directions in which the function is increasing and decreasing most rapidly.

Activity 10.18.

In this activity we investigate how the gradient is related to the directions of greatest increase and decrease of a function. Let f be a differentiable function and $\mathbf{u}$ a unit vector.

(a) Let θ be the angle between $\nabla f(x_0, y_0)$ and $\mathbf{u}$. Explain why

$$D_{\mathbf{u}}f(x_0, y_0) = |\langle f_x(x_0, y_0), f_y(x_0, y_0)\rangle|\cos(\theta). \tag{10.13}$$

(b) At the point (x_0, y_0), the only quantity in Equation (10.13) that can change is θ (which determines the direction **u** of travel). Explain why $\theta = 0$ makes the quantity

$$|\langle f_x(x_0, y_0), f_y(x_0, y_0)\rangle| \cos(\theta)$$

as large as possible.

(c) When $\theta = 0$, in what direction does the unit vector **u** point relative to $\nabla f(x_0, y_0)$? Why? What does this tell us about the direction of greatest increase of f at the point (x_0, y_0)?

(d) In what direction, relative to $\nabla f(x_0, y_0)$, does f decrease most rapidly at the point (x_0, y_0)?

(e) State the unit vectors **u** and **v** (in terms of $\nabla f(x_0, y_0)$) that provide the directions of greatest increase and decrease for the function f at the point (x_0, y_0). What important assumption must we make regarding $\nabla f(x_0, y_0)$ in order for these vectors to exist?

◁

The Length of the Gradient

Having established in Activity 10.18 that the direction in which a function increases most rapidly at a point (x_0, y_0) is the unit vector **u** in the direction of the gradient, (that is, $\mathbf{u} = \frac{1}{|\nabla f(x_0, y_0)|} \nabla f(x_0, y_0)$, provided that $\nabla f(x_0, y_0) \neq 0$), it is also natural to ask, "in the direction of greatest increase for f at (x_0, y_0), what is the *value* of the rate of increase?" In this situation, we are asking for the value of $D_{\mathbf{u}} f(x_0, y_0)$ where $\mathbf{u} = \frac{1}{|\nabla f(x_0, y_0)|} \nabla f(x_0, y_0)$.

Using the now familiar way to compute the directional derivative, we see that

$$D_{\mathbf{u}} f(x_0, y_0) = \nabla f(x_0, y_0) \cdot \left(\frac{1}{|\nabla f(x_0, y_0)|} \nabla f(x_0, y_0) \right).$$

Next, we recall two important facts about the dot product: (i) $\mathbf{w} \cdot (c\mathbf{v}) = c(\mathbf{w} \cdot \mathbf{v})$ for any scalar c, and (ii) $\mathbf{w} \cdot \mathbf{w} = |\mathbf{w}|^2$. Applying these properties to the most recent equation involving the directional derivative, we find that

$$D_{\mathbf{u}} f(x_0, y_0) = \frac{1}{|\nabla f(x_0, y_0)|} (\nabla f(x_0, y_0) \cdot \nabla f(x_0, y_0)) = \frac{1}{|\nabla f(x_0, y_0)|} |\nabla f(x_0, y_0)|^2.$$

Finally, since $\nabla f(x_0, y_0)$ is a nonzero vector, its length $|\nabla f(x_0, y_0)|$ is a nonzero scalar, and thus we can simplify the preceding equation to establish that

$$D_{\mathbf{u}} f(x_0, y_0) = |\nabla f(x_0, y_0)|.$$

We summarize our most recent work by stating two important facts about the gradient.

> Let f be a differentiable function and (x_0, y_0) a point for which $\nabla f(x_0, y_0) \neq 0$. Then $\nabla f(x_0, y_0)$ points in the direction of greatest increase of f at (x_0, y_0), and the instantaneous rate of change of f in that direction is the length of the gradient vector. That is, if $\mathbf{u} = \frac{1}{|\nabla f(x_0, y_0)|} \nabla f(x_0, y_0)$, then **u** is a unit vector in the direction of greatest increase of f at (x_0, y_0), and $D_{\mathbf{u}} f(x_0, y_0) = |\nabla f(x_0, y_0)|$.

Activity 10.19.

Consider the function f defined by $f(x, y) = 2x^2 - xy + 2y$.

(a) Find the gradient $\nabla f(1, 2)$ and sketch it on Figure 10.43.

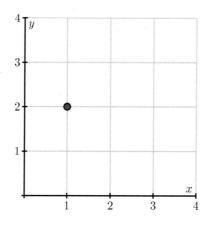

Figure 10.43: A plot for the gradient $\nabla f(1, 2)$.

(b) Sketch the unit vector $\mathbf{z} = \left\langle -\frac{1}{\sqrt{2}}, -\frac{1}{\sqrt{2}} \right\rangle$ on Figure 10.43 with its tail at $(1, 2)$. Now find the directional derivative $D_{\mathbf{z}}f(1, 2)$.

(c) What is the slope of the graph of f in the direction $\mathbf{z}$? What does the sign of the directional derivative tell you?

(d) Consider the vector $\mathbf{v} = \langle 2, -1 \rangle$ and sketch $\mathbf{v}$ on Figure 10.43 with its tail at $(1, 2)$. Find a unit vector $\mathbf{w}$ pointing in the same direction of $\mathbf{v}$. Without computing $D_{\mathbf{w}}f(1, 2)$, what do you know about the sign of this directional derivative? Now verify your observation by computing $D_{\mathbf{w}}f(1, 2)$.

(e) In which direction (that is, for what unit vector $\mathbf{u}$) is $D_{\mathbf{u}}f(1, 2)$ the greatest? What is the slope of the graph in this direction?

(f) Corresponding, in which direction is $D_{\mathbf{u}}f(1, 2)$ least? What is the slope of the graph in this direction?

(g) Sketch two unit vectors $\mathbf{u}$ for which $D_{\mathbf{u}}f(1, 2) = 0$ and then find component representations of these vectors.

(h) Suppose you are standing at the point $(3, 3)$. In which direction should you move to cause f to increase as rapidly as possible? At what rate does f increase in this direction?

◁

Applications

The gradient finds many natural applications. For example, situations often arise – for instance, constructing a road through the mountains or planning the flow of water across a landscape – where we are interested in knowing the direction in which a function is increasing or decreasing most rapidly.

For example, consider a two-dimensional version of how a heat-seeking missile might work.[5] Suppose that the temperature surrounding a fighter jet can be modeled by the function T defined by

$$T(x,y) = \frac{100}{1 + (x-5)^2 + 4(y-2.5)^2},$$

where (x,y) is a point in the plane of the fighter jet and $T(x,y)$ is measured in degrees Celsius. Some contours and gradients ∇T are shown on the left in Figure 10.44.

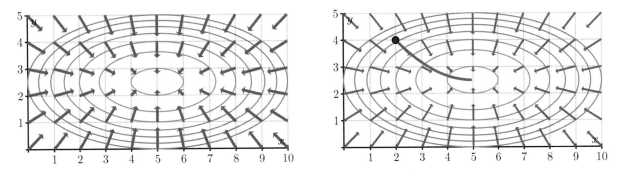

Figure 10.44: Contours and gradient for $T(x,y)$ and the missile's path.

A heat-seeking missile will always travel in the direction in which the temperature increases most rapidly; that is, it will always travel in the direction of the gradient ∇T. If a missile is fired from the point $(2,4)$, then its path will be that shown on the right in Figure 10.44.

In the final activity of this section, we consider several questions related to this context of a heat-seeking missile, and foreshadow some upcoming work in Section 10.7.

Activity 10.20.

(a) The temperature $T(x,y)$ has its maximum value at the fighter jet's location. State the fighter jet's location and explain how Figure 10.44 tells you this.

(b) Determine ∇T at the fighter jet's location and give a justification for your response.

(c) Suppose that a different function f has a local maximum value at (x_0, y_0). Sketch the behavior of some possible contours near this point. What is $\nabla f(x_0, y_0)$?

(d) Suppose that a function g has a local minimum value at (x_0, y_0). Sketch the behavior of some possible contours near this point. What is $\nabla g(x_0, y_0)$?

---

[5] This application is borrowed from United States Air Force Academy Department of Mathematical Sciences at http://www.nku.edu/~longa/classes/mat320/mathematica/multcalc.htm.

(e) If a function g has a local minimum at (x_0, y_0), what is the direction of greatest increase of g at (x_0, y_0)?

◁

Summary

- The directional derivative of f at the point (x, y) in the direction of the unit vector $\mathbf{u} = \langle u_1, u_2 \rangle$ is
$$D_{\mathbf{u}}f(x, y) = \lim_{h \to 0} \frac{f(x + u_1 h, y + u_2 h) - f(x, y)}{h}$$
for those values of x and y for which the limit exists. In addition, $D_{\mathbf{u}}f(x, y)$ measures the slope of the graph of f when we move in the direction $\mathbf{u}$. Alternatively, $D_{\mathbf{u}}f(x_0, y_0)$ measures the instantaneous rate of change of f in the direction $\mathbf{u}$ at (x_0, y_0).

- The gradient of a function $f = f(x, y)$ at a point (x_0, y_0) is the vector
$$\nabla f(x_0, y_0) = \langle f_x(x_0, y_0), f_y(x_0, y_0) \rangle.$$

- The directional derivative in the direction $\mathbf{u}$ may be computed by
$$D_{\mathbf{u}}f(x_0, y_0) = \nabla f(x_0, y_0) \cdot \mathbf{u}.$$

- At any point where the gradient is nonzero, gradient is orthogonal to the contour through that point and points in the direction in which f increases most rapidly; moreover, the slope of f in this direction equals the length of the gradient $|\nabla f(x_0, y_0)|$. Similarly, the opposite of the gradient points in the direction of greatest decrease, and that rate of decrease is the opposite of the length of the gradient.

Exercises

1. Let $E(x, y) = \dfrac{100}{1 + (x - 5)^2 + 4(y - 2.5)^2}$ represent the elevation on a land mass at location (x, y). Suppose that E, x, and y are all measured in meters.

   (a) Find $E_x(x, y)$ and $E_y(x, y)$.

   (b) Let $\mathbf{u}$ be a unit vector in the direction of $\langle -4, 3 \rangle$. Determine $D_{\mathbf{u}}E(3, 4)$. What is the practical meaning of $D_{\mathbf{u}}E(3, 4)$ and what are its units?

   (c) Find the direction of greatest increase in E at the point $(3, 4)$.

   (d) Find the instantaneous rate of change of E in the direction of greatest decrease at the point $(3, 4)$. Include units on your answer.

   (e) At the point $(3, 4)$, find a direction $\mathbf{w}$ in which the instantaneous rate of change of E is 0.

2. Let $f(x, y) = x^2 + 3y^2$.

(a) Find $\nabla f(x, y)$ and $\nabla f(1, 2)$.

(b) Find the direction of greatest increase in f at the point $(1, 2)$. Explain. A graph of the surface defined by f is shown in Figure 10.45. Illustrate this direction on the surface.

(c) A contour diagram of f is shown in Figure 10.46. Illustrate your calculation from (b) on this contour diagram.

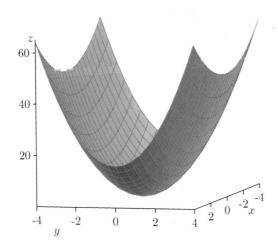

Figure 10.45: The surface for $f(x, y) = x^2 + 3y^2$.

Figure 10.46: Contours for $f(x, y) = x^2 + 3y^2$.

(d) Find a direction **w** for which the slope of the tangent line to the surface generated by f at the point $(1, 2)$ is zero in the direction **w**.

3. The properties of the gradient that we have observed for functions of two variables also hold for functions of more variables. In this problem, we consider a situation where there are three independent variables. Suppose that the temperature in a region of space is described by

$$T(x, y, z) = 100e^{-x^2-y^2-z^2}$$

and that you are standing at the point $(1, 2, -1)$.

(a) Find the instantaneous rate of change of the temperature in the direction of $\mathbf{v} = \langle 0, 1, 2 \rangle$ at the point $(1, 2, -1)$. Remember that you should first find a *unit* vector in the direction of **v**.

(b) In what direction from the point $(1, 2, -1)$ would you move to cause the temperature to decrease as quickly as possible?

(c) How fast does the temperature decrease in this direction?

(d) Find a direction in which the temperature does not change at $(1, 2, -1)$.

10.6. DIRECTIONAL DERIVATIVES AND THE GRADIENT

should be 0
∴ the z direction doesn't change.

contour!

90°

Contour lines should be always orthogonal to the gradient.

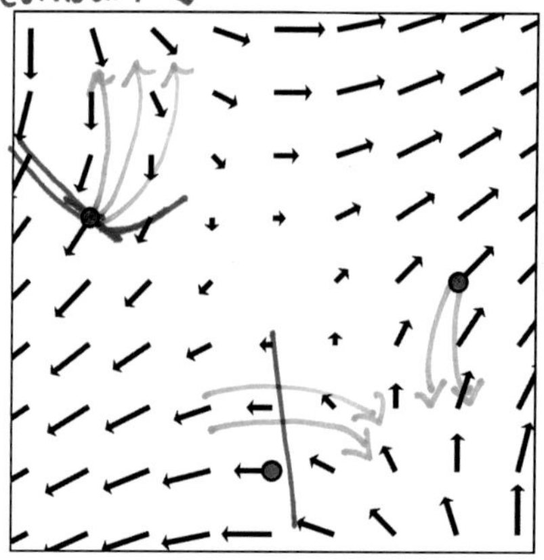

Figure 10.47: The gradient ∇f.

4. Figure 10.47 shows a plot of the gradient ∇f at several points for some function $f = f(x, y)$.

(a) Consider each of the three indicated points, and draw, as best as you can, the contour through that point.

(b) Beginning at each point, draw a curve on which f is continually decreasing.

10.7 Optimization

> **Motivating Questions**
>
> In this section, we strive to understand the ideas generated by the following important questions:
>
> - How can we find the points at which $f(x, y)$ has a local maximum or minimum?
> - How can we determine whether critical points of $f(x, y)$ are local maxima or minima?
> - How can we find the absolute maximum and minimum of $f(x, y)$ on a closed and bounded domain?

When is our func. big or small?

Introduction

We learn in single-variable calculus that the derivative is a useful tool for finding the local maxima and minima of functions, and that these ideas may often be employed in applied settings. In particular, if a function f, such as the one shown in Figure 10.48 is everywhere differentiable, we know that the tangent line is horizontal at any point where f has a local maximum or minimum. This, of course, means that the derivative f' is zero at any such point. Hence, one way that we seek extreme values of a given function is to first find where the derivative of the function is zero.

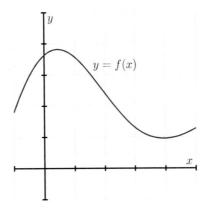

Figure 10.48: The graph of $y = f(x)$.

In multivariable calculus, we are often similarly interested in finding the greatest and/or least value(s) that a function may achieve. Moreover, there are many applied settings in which a quantity of interest depends on several different variables. In the following preview activity, we begin to see how some key ideas in multivariable calculus can help us answer such questions by thinking about the geometry of the surface generated by a function of two variables.

Preview Activity 10.7. Let $z = f(x, y)$ be a differentiable function, and suppose that at the point (x_0, y_0), f achieves a local maximum. That is, the value of $f(x_0, y_0)$ is greater than the value of $f(x,y)$ for all (x,y) nearby (x_0, y_0). You might find it helpful to sketch a rough picture of a possible function f that has this property.

(a) If we consider the trace given by holding $y = y_0$ constant, then the single-variable function defined by $f(x, y_0)$ must have a local maximum at x_0. What does this say about the value of the partial derivative $f_x(x_0, y_0)$?

(b) In the same way, the trace given by holding $x = x_0$ constant has a local maximum at $y = y_0$. What does this say about the value of the partial derivative $f_y(x_0, y_0)$?

(c) What may we now conclude about the gradient $\nabla f(x_0, y_0)$ at the local maximum? How is this consistent with the statement "f increases most rapidly in the direction $\nabla f(x_0, y_0)$?"

(d) How will the tangent plane to the surface $z = f(x, y)$ appear at the point $(x_0, y_0, f(x_0, y_0))$?

(e) By first computing the partial derivatives, find any points at which $f(x, y) = 2x - x^2 - (y+2)^2$ may have a local maximum.

⋈

Extrema and Critical Points

One of the important applications of single-variable calculus is the use of derivatives to identify local extremes of functions (that is, local maxima and local minima). Using the tools we have developed so far, we can naturally extend the concept of local maxima and minima to several-variable functions.

> **Definition 10.5.** Let f be a function of two variables x and y.
>
> - The function f has a **local maximum** at a point (x_0, y_0) provided that $f(x, y) \leq f(x_0, y_0)$ for all points (x, y) near (x_0, y_0). In this situation we say that $f(x_0, y_0)$ is a **local maximum value**.
>
> - The function f has a **local minimum** at a point (x_0, y_0) provided that $f(x, y) \geq f(x_0, y_0)$ for all points (x, y) near (x_0, y_0). In this situation we say that $f(x_0, y_0)$ is a **local minimum value**.
>
> - An **absolute maximum point** is a point (x_0, y_0) for which $f(x, y) \leq f(x_0, y_0)$ for all points (x, y) in the domain of f. The value of f at an absolute maximum point is the **maximum value** of f.
>
> - An **absolute minimum point** is a point such that $f(x, y) \geq f(x_0, y_0)$ for all points (x, y) in the domain of f. The value of f at an absolute minimum point is the **maximum value** of f.

10.7. OPTIMIZATION

We use the term **extremum point** to refer to any point (x_0, y_0) at which f has a local maximum or minimum. In addition, the function value $f(x_0, y_0)$ at an extremum is called an **extremal value**. Figure 10.49 illustrates the graphs of two functions that have an absolute maximum and minimum, respectively, at the origin $(x_0, y_0) = (0, 0)$.

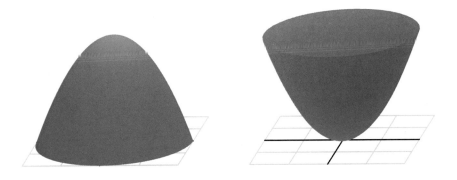

Figure 10.49: An absolute maximum and an absolute minimum

In single-variable calculus, we saw that the extrema of a continuous function f always occur at *critical points*, values of x where f fails to be differentiable or where $f'(x) = 0$. Said differently, critical points provide the locations where extrema of a function may appear. Our work in Preview Activity 10.7 suggests that something similar happens with two-variable functions.

Suppose that a continuous function f has an extremum at (x_0, y_0). In this case, the trace $f(x, y_0)$ has an extremum at x_0, which means that x_0 is a critical value of $f(x, y_0)$. Therefore, either $f_x(x_0, y_0)$ does not exist or $f_x(x_0, y_0) = 0$. Similarly, either $f_y(x_0, y_0)$ does not exist or $f_y(x_0, y_0) = 0$. This implies that the extrema of a two-variable function occur at points that satisfy the following definition.

> **Definition 10.6.** A **critical point** (x_0, y_0) of a function $f = f(x, y)$ is a point in the domain of f at which $f_x(x_0, y_0) = 0$ and $f_y(x_0, y_0) = 0$, or such that one of $f_x(x_0, y_0)$ or $f_y(x_0, y_0)$ fails to exist.

We can therefore find critical points of a function f by computing partial derivatives and identifying any values of (x, y) for which one of the partials doesn't exist or for which both partial derivatives are simultaneously zero. For the latter, note that we have to solve the system of equations

$$f_x(x, y) = 0$$
$$f_y(x, y) = 0.$$

Activity 10.21.

Find the critical points of each of the following functions. Then, using appropriate technology (e.g., Wolfram|Alpha or CalcPlot3D[6]), plot the graphs of the surfaces near each critical value and compare the graph to your work.

(a) $f(x, y) = 2 + x^2 + y^2$

(b) $f(x, y) = 2 + x^2 - y^2$

(c) $f(x, y) = 2x - x^2 - \frac{1}{4}y^2$

(d) $f(x, y) = |x| + |y|$

(e) $f(x, y) = 2xy - 4x + 2y - 3$.

◁

Classifying Critical Points: The Second Derivative Test

While the extrema of a continuous function f always occur at critical points, it is important to note that not every critical point leads to an extremum. Recall, for instance, $f(x) = x^3$ from single variable calculus. We know that $x_0 = 0$ is a critical point since $f'(x_0) = 0$, but $x_0 = 0$ is neither a local maximum nor a local minimum of f.

A similar situation may arise in a multivariable setting. Consider the function f defined by $f(x, y) = x^2 - y^2$ whose graph and contour plot are shown in Figure 10.50. Because $\nabla f = \langle 2x, -2y \rangle$, we see that the origin $(x_0, y_0) = (0, 0)$ is a critical point. However, this critical point is neither a local maximum or minimum; the origin is a local minimum on the trace defined by $y = 0$, while the origin is a local maximum on the trace defined by $x = 0$. We call such a critical point a **saddle point** due to the shape of the graph near the critical point.

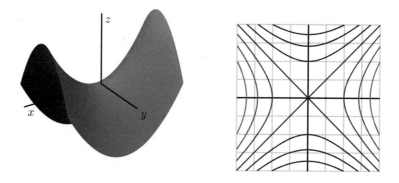

Figure 10.50: A saddle point.

---

[6]at http://web.monroecc.edu/manila/webfiles/calcNSF/JavaCode/CalcPlot3D.htm

10.7. OPTIMIZATION

As in single-variable calculus, we would like to have some sort of test to help us identify whether a critical point is a local maximum, local maximum, or neither. Before describing the test that accomplishes this classification in a two-variable setting, we recall the Second Derivative Test from single-variable calculus.

> *of many 2nd order second derivatives!*
>
> **The Second Derivative Test** for single-variable functions. If x_0 is a critical point of a function f so that $f'(x_0) = 0$ and if $f''(x_0)$ exists, then
>
> - if $f''(x_0) < 0$, x_0 is a local maximum,
> - if $f''(x_0) > 0$, x_0 is a local minimum, and
> - if $f''(x_0) = 0$, this test yields no information.

For the analogous test for functions of two variables, we have to consider all of the second-order partial derivatives.

> **The Second Derivative Test** for two-variable functions. Suppose (x_0, y_0) is a critical point of the function f for which $f_x(x_0, y_0) = 0$ and $f_y(x_0, y_0) = 0$. Let D be the quantity defined by
>
> $$D = f_{xx}(x_0, y_0) f_{yy}(x_0, y_0) - f_{xy}(x_0, y_0)^2.$$
>
> *must have same sign*
>
> - If $D > 0$ and $f_{xx}(x_0, y_0) < 0$, then f has a local maximum at (x_0, y_0).
> - If $D > 0$ and $f_{xx}(x_0, y_0) > 0$, then f has a local minimum at (x_0, y_0).
> - If $D < 0$, then f has a saddle point at (x_0, y_0).
> - If $D = 0$, then this test yields no information about what happens at (x_0, y_0).
>
> The quantity D is called the *discriminant* of the function f at (x_0, y_0).

To properly understand the origin of the Second Derivative Test, we could introduce a "second-order directional derivative." If this second-order directional derivative were negative in every direction, for instance, we could guarantee that the critical point is a local maximum. A complete justification of the Second Derivative Test requires key ideas from linear algebra that are beyond the scope of this course, so instead of presenting a detailed explanation, we will accept this test as stated and demonstrate its use in three basic examples.

Example 10.1. Let $f(x,y) = 4 - x^2 - y^2$ as shown in Figure 10.51. Critical points occur when $f_x = -2x = 0$ and $f_y = -2y = 0$ so the origin $(x_0, y_0) = (0,0)$ is the only critical point. We then find that

$$f_{xx}(0,0) = -2, f_{yy}(0,0) = -2, \text{ and } f_{xy}(0,0) = 0,$$

giving $D = (-2)(-2) - 0^2 = 4 > 0$. We then consider $f_{xx}(0,0) = -2 < 0$ and conclude, from the Second Derivative Test, that $f(0,0) = 4$ is a local maximum value.

Figure 10.51: $z = 4 - x^2 - y^2$

Example 10.2. Let $f(x,y) = x^2 + y^2$ as shown in Figure 10.52. Critical points occur when $f_x = 2x = 0$ and $f_y = 2y = 0$ so the origin $(x_0, y_0) = (0,0)$ is the only critical point. We then find that

$$f_{xx}(0,0) = 2, f_{yy}(0,0) = 2, \text{ and } f_{xy}(0,0) = 0,$$

giving $D = 2 \cdot 2 - 0^2 = 4 > 0$. We then consider $f_{xx}(0,0) = 2 > 0$ and conclude, from the Second Derivative Test, that $f(0,0) = 0$ is a local minimum value.

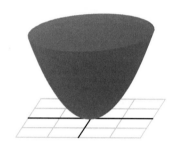

Figure 10.52: $z = x^2 + y^2$

Example 10.3. Let $f(x,y) = x^2 - y^2$ as shown in Figure 10.53. Critical points occur when $f_x = 2x = 0$ and $f_y = -2y = 0$ so the origin $(x_0, y_0) = (0,0)$ is the only critical point. We then find that

$$f_{xx}(0,0) = 2, f_{yy}(0,0) = -2, \text{ and } f_{xy}(0,0) = 0,$$

giving $D = 2 \cdot (-2) - 0^2 = -4 < 0$. We then conclude, from the Second Derivative Test, that $(0,0)$ is a saddle point.

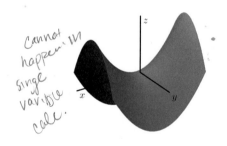

Cannot happen since single variable calc.

Figure 10.53: $z = x^2 - y^2$

Activity 10.22.

Find the critical points of the following functions and use the Second Derivative Test to classify the critical points.

(a) $f(x,y) = 3x^3 + y^2 - 9x + 4y$

(b) $f(x,y) = xy + \frac{2}{x} + \frac{4}{y}$

(c) $f(x,y) = x^3 + y^3 - 3xy$.

◁

As we learned in single-variable calculus, finding extremal values of functions can be particularly useful in applied settings. For instance, we can often use calculus to determine the least

10.7. OPTIMIZATION

expensive way to construct something or to find the most efficient route between two locations. The same possibility holds in settings with two or more variables.

Activity 10.23.

While the quantity of a product demanded by consumers is often a function of the price of the product, the demand for a product may also depend on the price of other products. For instance, the demand for blue jeans at Old Navy may be affected not only by the price of the jeans themselves, but also by the price of khakis.

Suppose we have two goods whose respective prices are p_1 and p_2. The demand for these goods, q_1 and q_2, depend on the prices as

$$q_1 = 150 - 2p_1 - p_2 \qquad (10.14)$$
$$q_2 = 200 - p_1 - 3p_2. \qquad (10.15)$$

The seller would like to set the prices p_1 and p_2 in order to maximize revenue. We will assume that the seller meets the full demand for each product. Thus, if we let R be the revenue obtained by selling q_1 items of the first good at price p_1 per item and q_2 items of the second good at price p_2 per item, we have

$$R = p_1 q_1 + p_2 q_2.$$

We can then write the revenue as a function of just the two variables p_1 and p_2 by using Equations (10.14) and (10.15), giving us

$$R(p_1, p_2) = p_1(150 - 2p_1 - p_2) + p_2(200 - p_1 - 3p_2) = 150p_1 + 200p_2 - 2p_1 p_2 - 2p_1^2 - 3p_2^2.$$

A graph of R as a function of p_1 and p_2 is shown in Figure 10.54.

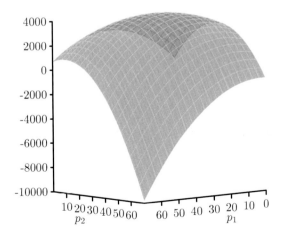

Figure 10.54: A revenue function.

(a) Find all critical points of the revenue function, R.

(b) Apply the Second Derivative Test to determine the type of any critical points.

(c) Where should the seller set the prices p_1 and p_2 to maximize the revenue?

◁

Optimization on a Restricted Domain

The Second Derivative Test helps us classify critical points of a function, but it does not tell us if the function actually has an absolute maximum or minimum at each such point. For single-variable functions, the Extreme Value Theorem told us that a continuous function on a closed interval $[a, b]$ always has both an absolute maximum and minimum on that interval, and that these absolute extremes must occur at either an endpoint or at a critical point. Thus, to find the absolute maximum and minimum, we determine the critical points in the interval and then evaluate the function at these critical point s and at the endpoints of the interval. A similar approach works for functions of two variables.

For functions of two variables, closed and bounded regions play the role that closed intervals did for functions of a single variable. A closed region is a region that contains its boundary (the unit disk $x^2 + y^2 \leq 1$ is closed, while its interior $x^2 + y^2 < 1$ is not, for example), while a bounded region is one that does not stretch to infinity in any direction. Just as for functions of a single variable, continuous functions of several variables that are defined on closed, bounded regions must have absolute maxima and minima in those regions.

The Extreme Value Theorem. Let $f = f(x, y)$ be a continuous function on a closed and bounded region R. Then f has an absolute maximum and an absolute minimum in R.

The absolute extremes must occur at either a critical point in the interior of R or at a boundary point of R. We therefore must test both possibilities, as we demonstrate in the following example.

Example 10.4. Suppose the temperature T at each point on the circular plate $x^2 + y^2 \leq 1$ is given by
$$T(x, y) = 2x^2 + y^2 - y.$$
The domain $R = \{(x, y) : x^2 + y^2 \leq 1\}$ is a closed and bounded region, as shown on the left of Figure 10.55, so the Extreme Value Theorem assures us that T has an absolute maximum and minimum on the plate. The graph of T over its domain R is shown in Figure 10.55. We will find the hottest and coldest points on the plate.

If the absolute maximum or minimum occurs inside the disk, it will be at a critical point so we begin by looking for critical points inside the disk. To do this, notice that critical points are given by the conditions $T_x = 4x = 0$ and $T_y = 2y - 1 = 0$. This means that there is one critical point of the function at the point $(x_0, y_0) = (0, 1/2)$, which lies inside the disk.

We now find the hottest and coldest points on the boundary of the disk, which is the circle of radius 1. As we have seen, the points on the unit circle can be parametrized as
$$x(t) = \cos(t), \ y(t) = \sin(t),$$

10.7. OPTIMIZATION

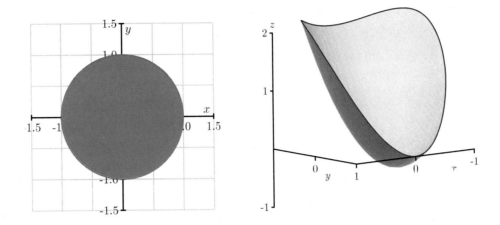

Figure 10.55: Domain of the temperature $T(x, y) = 2x^2 + y^2 - y$ and its graph.

where $0 \leq t \leq 2\pi$. The temperature at a point on the circle is then described by [parameterization]

$$T(x(t), y(t)) = 2\cos^2(t) + \sin^2(t) - \sin(t).$$

To find the hottest and coldest point on the boundary, we look for the critical points of this single-variable function on the interval $0 \leq t \leq 2\pi$. We have

$$\frac{dT}{dt} = -4\cos(t)\sin(t) + 2\cos(t)\sin(t) - \cos(t) = -2\cos(t)\sin(t) - \cos(t) = \cos(t)(-2\sin(t) - 1) = 0.$$

This shows that we have critical points when $\cos(t) = 0$ or $\sin(t) = -1/2$. This occurs when $t = \pi/2, 3\pi/2, 7\pi/6$, and $11\pi/6$. Since we have $x(t) = \cos(t)$ and $y(t) = \sin(t)$, the corresponding points are

- $(x, y) = (0, 1)$ when $t = \frac{\pi}{2}$,
- $(x, y) = (0, -1)$ when $t = \frac{3\pi}{2}$,
- $(x, y) = \left(-\frac{\sqrt{3}}{2}, -\frac{1}{2}\right)$ when $t = \frac{7\pi}{6}$,
- $(x, y) = \left(\frac{\sqrt{3}}{2}, -\frac{1}{2}\right)$ when $t = \frac{11\pi}{6}$.

These are the critical points of T on the boundary and so this collection of points includes the hottest and coldest points on the boundary.

We now have a list of candidates for the hottest and coldest points: the critical point in the interior of the disk and the critical points on the boundary. We find the hottest and coldest points by evaluating the temperature at each of these points, and find that

- $T\left(0, \frac{1}{2}\right) = -\frac{1}{4}$,
- $T\left(-\frac{\sqrt{3}}{2}, -\frac{1}{2}\right) = \frac{9}{4}$,
- $T(0, 1) = 0$,
- $T\left(\frac{\sqrt{3}}{2}, -\frac{1}{2}\right) = \frac{9}{4}$.
- $T(0, -1) = 2$,

So the maximum value of T on the disk $x^2 + y^2 \leq 1$ is $\frac{9}{4}$, which occurs at the two points $\left(\pm\frac{\sqrt{3}}{2}, -\frac{1}{2}\right)$ on the boundary, and the minimum value of T on the disk is $-\frac{1}{4}$ which occurs at the critical point $\left(0, \frac{1}{2}\right)$ in the interior of R.

From this example, we see that we use the following procedure for determining the absolute maximum and absolute minimum of a function on a closed and bounded domain.

Step 1: Find all critical points of the function in the interior of the domain.

Step 2: Find all the critical points of the function on the boundary of the domain. Working on the boundary of the domain reduces this part of the problem to one or more single variable optimization problems.

Step 3: Evaluate the function at each of the points found in Steps 1 and 2.

Step 4: The maximum value of the function is the largest value obtained in Step 3, and the minimum value of the function is the smallest value obtained in Step 3.

Activity 10.24.

Let $f(x, y) = x^2 - 3y^2 - 4x + 6y$ with triangular domain R whose vertices are at $(0, 0)$, $(4, 0)$, and $(0, 4)$. The domain R and a graph of f on the domain appear in Figure 10.56.

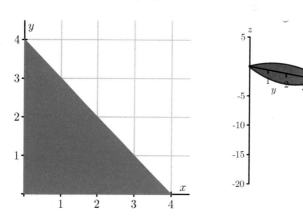

Figure 10.56: The domain of $f(x, y) = x^2 - 3y^2 - 4x + 6y$ and its graph.

(a) Find all of the critical points of f in R.

(b) Parameterize the horizontal leg of the triangular domain, and find the critical points of f on that leg.

(c) Parameterize the vertical leg of the triangular domain, and find the critical points of f on that leg.

(d) Parameterize the hypotenuse of the triangular domain, and find the critical points of f on the hypotenuse.

10.7. OPTIMIZATION

(e) Find the absolute maximum and absolute minimum value of f on R.

◁

Summary

- To find the extrema of a function $f = f(x,y)$, we first find the critical points, which are points where one of the partials of f fails to exist, or where $f_x = 0$ and $f_y = 0$.

- The Second Derivative Test helps determine whether a critical point is a local maximum, local minimum, or saddle point.

- If f is defined on a closed and bounded domain, we find the absolute maxima and minima by finding the critical points in the interior of the domain, finding the critical points on the boundary, and testing the value of f at both sets of critical points.

Exercises

1. Respond to each of the following prompts to solve the given optimization problem.

    (a) Let $f(x,y) = \sin(x) + \cos(y)$. Determine the absolute maximum and minimum values of f. At what points do these extreme values occur?

    (b) For a certain differentiable function F of two variables x and y, its partial derivatives are
    $$F_x(x,y) = x^2 - y - 4 \quad \text{and} \quad F_y(x,y) = -x + y - 2.$$
    Find each of the critical points of F, and classify each as a local maximum, local minimum, or a saddle point.

    (c) Determine all critical points of $T(x,y) = 48 + 3xy - x^2 y - xy^2$ and classify each as a local maximum, local minimum, or saddle point.

    (d) Find and classify all critical points of $g(x,y) = \frac{x^2}{2} + 3y^3 + 9y^2 - 3xy + 9y - 9x$.

    (e) Find and classify all critical points of $z = f(x,y) = ye^{-x^2 - 2y^2}$.

    (f) Determine the absolute maximum and absolute minimum of $f(x,y) = 2 + 2x + 2y - x^2 - y^2$ on the triangular plate in the first quadrant bounded by the lines $x = 0$, $y = 0$, and $y = 9 - x$.

    (g) Determine the absolute maximum and absolute minimum of $f(x,y) = 2 + 2x + 2y - x^2 - y^2$ over the closed disk of points (x,y) such that $(x-1)^2 + (y-1)^2 \leq 1$.

    (h) Find the point on the plane $z = 6 - 3x - 2y$ that lies closest to the origin.

2. If a continuous function f of a single variable has two critical numbers c_1 and c_2 at which f has relative maximum values, then f must have another critical number c_3, because "it is impossible to have two mountains without some sort of valley in between. The other critical point can be a saddle point (a pass between the mountains) or a local minimum (a true valley)."[7]

---

[7]From *Calculus in Vector Spaces* by Lawrence J. Corwin and Robert H. Szczarb.

10.7. OPTIMIZATION

Consider the function f defined by $f(x,y) = 4x^2 e^y - 2x^4 - e^{4y}$.[8] Show that f has exactly two critical points of f, and that f has relative maximum values at each of these critical points. Explain how function illustrates that it really is possible to have two mountains without some sort of valley in between. Use appropriate technology to draw the surface defined by f to see graphically how this happens.

3. If a continuous function f of a single variable has exactly one critical number with a relative maximum at that critical point, then the value of f at that critical point is an absolute maximum. In this exercise we see that the same is not always true for functions of two variables. Let $f(x,y) = 3xe^y - x^3 - e^{3y}$[9] Show that f has exactly one critical point, has a relative maximum value at that critical point, but that f has no absolute maximum value. Use appropriate technology to draw the surface defined by f to see graphically how this happens.

4. A manufacturer wants to procure rectangular boxes to ship its product. The boxes must contain 20 cubic feet of space. To be durable enough to ensure the safety of the product, the material for the sides of the boxes will cost $0.10 per square foot, while the material for the top and bottom will cost $0.25 per square foot. In this activity we will help the manufacturer determine the box of minimal cost.

    (a) What quantities are constant in this problem? What are the variables in this problem? Provide appropriate variable labels. What, if any, restrictions are there on the variables?

    (b) Using your variables from (a), determine a formula for the total cost C of a box.

    (c) Your formula in part (b) might be in terms of three variables. If so, find a relationship between the variables, and then use this relationship to write C as a function of only two independent variables.

    (d) Find the dimensions that minimize the cost of a box. Be sure to verify that you have a minimum cost.

5. A rectangular box with length x, width y, and height z is being built. The box is positioned so that one corner is stationed at the origin and the box lies in the first octant where x, y, and z are all positive. There is an added constraint on how the box is constructed: it must fit underneath the plane with equation $x + 2y + 3z = 6$. In fact, we will assume that the corner of the box "opposite" the origin must actually lie on this plane. The basic problem is to find the maximum volume of the box.

    (a) Sketch the plane $x + 2y + 3z = 6$, as well as a picture of a potential box. Label everything appropriately.

    (b) Explain how you can use the fact that one corner of the box lies on the plane to write the volume of the box as a function of x and y only. Do so, and clearly show the formula you find for $V(x, y)$.

---

[8] From Ira Rosenholz in the Problems Section of the *Mathematics Magazine*, Vol. 60 NO. 1, February 1987.

[9] From "'The Only Critical Point in Town' Test' by Ira Rosenholz and Lowell Smylie in the *Mathematics Magazine*, VOL 58 NO 3 May 1985.

(c) Find all critical points of V. (Note that when finding the critical points, it is essential that you factor first to make the algebra easier.)

(d) Without considering the current applied nature of the function V, classify each critical point you found above as a local maximum, local minimum, or saddle point of V.

(e) Determine the maximum volume of the box, justifying your answer completely with an appropriate discussion of the critical points of the function.

(f) Now suppose that we instead stipulated that, while the vertex of the box opposite the origin still had to lie on the plane, we were only going to permit the sides of the box, x and y, to have values in a specified range (given below). That is, we now want to find the maximum value of V on the closed, bounded region

$$\frac{1}{2} \leq x \leq 1, \ 1 \leq y \leq 2.$$

Find the maximum volume of the box under this condition, justifying your answer fully.

6. The airlines place restrictions on luggage that can be carried onto planes.

- A carry-on bag can weigh no more than 40 lbs.
- The length plus width plus height of a bag cannot exceed 45 inches.
- The bag must fit in an overhead bin.

Let x, y, and z be the length, width, and height (in inches) of a carry on bag. In this problem we find the dimensions of the bag of largest volume, $V = xyz$, that satisfies the second restriction. Assume that we use all 45 inches to get a maximum volume. (Note that this bag of maximum volume might not satisfy the third restriction.)

(a) Write the volume $V = V(x, y)$ as a function of just the two variables x and y.

(b) Explain why the domain over which V is defined is the triangular region R with vertices (0,0), (45,0), and (0,45).

(c) Find the critical points, if any, of V in the interior of the region R.

(d) Find the maximum value of V on the boundary of the region R, and the determine the dimensions of a bag with maximum volume on the entire region R. (Note that most carry-on bags sold today measure 22 by 14 by 9 inches with a volume of 2772 cubic inches, so that the bags will fit into the overhead bins.)

7. According to *The Song of Insects* by G.W. Pierce (Harvard College Press, 1948) the sound of striped ground crickets chirping, in number of chirps per second, is related to the temperature. So the number of chirps per second could be a predictor of temperature. The data Pierce collected is shown in the table below, where x is the (average) number of chirps per second and y is the temperature in degrees Fahrenheit.

10.7. OPTIMIZATION

x	y
20.0	88.6
16.0	71.6
19.8	93.3
18.4	84.3
17.1	80.6
15.5	75.2
14.7	69.7
17.1	82.0
15.4	69.4
16.2	83.3
15.0	79.6
17.2	82.6
16.0	80.6
17.0	83.5
14.4	76.3

A scatter plot of the data is given below. The relationship between x and y is not exactly linear, but looks to have a linear pattern. It could be that the relationship is really linear but experimental error causes the data to be slightly inaccurate. Or perhaps the data is not linear, but only approximately linear.

If we want to use the data to make predications, then we need to fit a curve of some kind to the data. Since the cricket data appears roughly linear, we will fit a linear function f of the form $f(x) = mx + b$ to the data. We will do this in such a way that we minimize the sums of the squares of the distances between the y values of the data and the corresponding y values of the line defined by f. This type of fit is called a *least squares* approximation. If the data is represented by the points $(x_1, y_1), (x_2, y_2), \ldots, (x_n, y_n)$, then the square of the distance between y_i and $f(x_i)$ is $(f(x_i) - y_i)^2 = (mx_i + b - y_i)^2$. So our goal is to minimize the sum of these squares, of minimize the function S defined by

$$S(m, b) = \sum_{i=1}^{n} (mx_i + b - y_i)^2.$$

(a) Calculate S_m and S_b.

(b) Solve the system $S_m(m, b) = 0$ and $S_b(m, b) = 0$ to show that the critical point satisfies

$$m = \frac{n\left(\sum_{i=1}^{n} x_i y_i\right) - \left(\sum_{i=1}^{n} x_i\right)\left(\sum_{i=1}^{n} y_i\right)}{n\left(\sum_{i=1}^{n} x_i^2\right) - \left(\sum_{i=1}^{n} x_i\right)^2}$$

$$b = \frac{\left(\sum_{i=1}^{n} y_i\right)\left(\sum_{i=1}^{n} x_i^2\right) - \left(\sum_{i=1}^{n} x_i\right)\left(\sum_{i=1}^{n} x_i y_i\right)}{n\left(\sum_{i=1}^{n} x_i^2\right) - \left(\sum_{i=1}^{n} x_i\right)^2}.$$

(Hint: Don't be daunted by these expressions, the system $S_m(m,b) = 0$ and $S_b(m,b) = 0$ is a system of two linear equations in the unknowns m and b. It might be easier to let $r = \sum_{i=1}^{n} x_i^2$, $s = \sum_{i=1}^{n} x_i$, $t = \sum_{i=1}^{n} y_i$, and $u = \sum_{i=1}^{n} x_i y_i$ and write your equations using these constants.)

(c) Use the Second Derivative Test to explain why the critical point gives a local minimum. Can you then explain why the critical point gives an absolute minimum?

(d) Use the formula from part (b) to find the values of m and b that give the line of best fit in the least squares sense to the cricket data. Draw your line on the scatter plot to convince yourself that you have a well-fitting line.

10.8 Constrained Optimization:Lagrange Multipliers

Motivating Questions

In this section, we strive to understand the ideas generated by the following important questions:

- What geometric condition enables us to optimize a function $f = f(x, y)$ subject to a constraint given by $g(x, y) = k$, where k is a constant?
- How can we exploit this geometric condition to find the extreme values of a function subject to a constraint?

Introduction

We previously considered how to find the extreme values of functions on both unrestricted domains and on closed, bounded domains. Other types of optimization problems involve maximizing or minimizing a quantity subject to an external constraint. In these cases the extreme values frequently won't occur at the points where the gradient is zero, but rather at other points that satisfy an important geometric condition. These problems are often called *constrained optimization* problems and can be solved with the method of Lagrange Multipliers, which we study in this section.

Preview Activity 10.8. According to U.S. postal regulations, the girth plus the length of a parcel sent by mail may not exceed 108 inches, where by "girth" we mean the perimeter of the smallest end. Our goal is to find the largest possible volume of a rectangular parcel with a square end that can be sent by mail.[10] If we let x be the length of the side of one square end of the package and y the length of the package, then we want to maximize the volume $f(x, y) = x^2 y$ of the box subject to the constraint that the girth ($4x$) plus the length (y) is as large as possible, or $4x + y = 108$. The equation $4x + y = 108$ is thus an external constraint on the variables.

(a) The constraint equation involves the function g that is given by

$$g(x, y) = 4x + y.$$

Explain why the constraint is a contour of g, and is therefore a two-dimensional curve.

(b) Figure 10.57 shows the graph of the constraint equation $g(x, y) = 108$ along with a few contours of the volume function f. Since our goal is to find the maximum value of f subject to the constraint $g(x, y) = 108$, we want to find the point on our constraint curve that intersects the contours of f at which f has its largest value.

---

[10]We solved this applied optimization problem in single variable *Active Calculus*, so it may look familiar. We take a different approach in this section, and this approach allows us to view most applied optimization problems from single variable calculus as constrained optimization problems, as well as provide us tools to solve a greater variety of optimization problems.

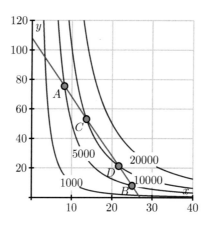

Figure 10.57: Contours of f and the constraint equation $g(x, y) = 108$.

- i. Points A and B in Figure 10.57 lie on a contour of f and on the constraint equation $g(x, y) = 108$. Explain why neither A nor B provides a maximum value of f that satisfies the constraint.

- ii. Points C and D in Figure 10.57 lie on a contour of f and on the constraint equation $g(x, y) = 108$. Explain why neither C nor D provides a maximum value of f that satisfies the constraint.

- iii. Based on your responses to parts i. and ii., draw the contour of f on which you believe f will achieve a maximum value subject to the constraint $g(x, y) = 108$. Explain why you drew the contour you did.

(c) Recall that $g(x, y) = 108$ is a contour of the function g, and that the gradient of a function is always orthogonal to its contours. With this in mind, how should ∇f and ∇g be related at the optimal point? Explain.

Constrained Optimization and Lagrange Multipliers

In Preview Activity 10.8, we considered an optimization problem where there is an external constraint on the variables, namely that the girth plus the length of the package cannot exceed 108 inches. We saw that we can create a function g from the constraint, specifically $g(x, y) = 4x + y$. The constraint equation is then just a contour of g, $g(x, y) = c$, where c is a constant (in our case 108). Figure 10.58 illustrates that the volume function f is maximized, subject to the constraint $g(x, y) = c$, when the graph of $g(x, y) = c$ is tangent to a contour of f. Moreover, the value of f on this contour is the sought maximum value. To find this point where the graph of the constraint is tangent to a contour of f, recall that ∇f is perpendicular to the contours of f and ∇g is perpendicular to the contour of g. At such a point, the vectors ∇g and ∇f are parallel, and thus we need

10.8. CONSTRAINED OPTIMIZATION: LAGRANGE MULTIPLIERS

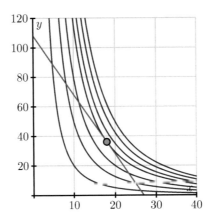

Figure 10.58: Contours of f and the constraint contour.

to determine the points where this occurs. Recall that two vectors are parallel if one is a nonzero scalar multiple of the other, so we therefore look for values of a parameter λ that make

$$\nabla f = \lambda \nabla g. \tag{10.16}$$

The constant λ is called a *Lagrange multiplier*.

To find the values of λ that satisfy (10.16) for the volume function in Preview Activity 10.8, we calculate both ∇f and ∇g. Observe that

$$\nabla f = 2xy\mathbf{i} + x^2\mathbf{j} \quad \text{and} \quad \nabla g = 4\mathbf{i} + \mathbf{j},$$

and thus we need a value of λ so that

$$2xy\mathbf{i} + x^2\mathbf{j} = \lambda(4\mathbf{i} + \mathbf{j}).$$

Equating components in the most recent equation and incorporating the original constraint, we have three equations

$$2xy = \lambda(4) \tag{10.17}$$
$$x^2 = \lambda(1) \tag{10.18}$$
$$4x + y = 108 \tag{10.19}$$

in the three unknowns x, y, and λ. First, note that if $\lambda = 0$, then equation (10.18) shows that $x = 0$. From this, Equation (10.19) tells us that $y = 108$. So the point $(0, 108)$ is a point we need to consider. Next, provided that $\lambda \neq 0$ (from which it follows that $x \neq 0$ by Equation (10.18)), we may divide both sides of Equation (10.17) by the corresponding sides of (10.18) to eliminate λ, and thus find that

$$\frac{2y}{x} = 4, \text{ so}$$
$$y = 2x.$$

Substituting into Equation (10.19) gives us

$$4x + 2x = 108$$

or

$$x = 18.$$

Thus we have $y = 2x = 36$ and $\lambda = x^2 = 324$ as another point to consider. So the points at which the gradients of f and g are parallel, and thus at which f may have a maximum or minimum subject to the constraint, are $(0, 108)$ and $(18, 36)$. By evaluating the function f at these points, we see that we maximize the volume when the length of the square end of the box is 18 inches and the length is 36 inches, for a maximum volume of $f(18, 36) = 11664$ cubic inches. Since $f(0, 108) = 0$, we obtain a minimum value at this point.

We summarize the process of Lagrange multipliers as follows.

> The general technique for optimizing a function $f = f(x, y)$ subject to a constraint $g(x, y) = c$ is to solve the system $\nabla f = \lambda \nabla g$ for x, y, and λ. We then evaluate the function f at each point (x, y) that results from a solution to the system in order to find the optimum values of f subject to the constraint.

Activity 10.25.

A cylindrical soda can holds about 355 cc of liquid. In this activity, we want to find the dimensions of such a can that will minimize the surface area.

(a) What are the variables in this problem? What restriction(s), if any, are there on these variables?

(b) What quantity do we want to optimize in this problem? What equation describes the constraint?

(c) Find λ and the values of your variables that satisfy Equation (10.16) in the context of this problem.

(d) Determine the dimensions of the pop can that give the desired solution to this constrained optimization problem.

The method of Lagrange multipliers also works for functions of more than two variables.

Activity 10.26.

Use the method of Lagrange multipliers to find the dimensions of the least expensive packing crate with a volume of 240 cubic feet when the material for the top costs $2 per square foot, the bottom is $3 per square foot and the sides are $1.50 per square foot.

Summary

- The extrema of a function $f = f(x, y)$ subject to a constraint $g(x, y) = c$ occur at points for which the contour of f is tangent to the curve that represents the constraint equation. This occurs when
$$\nabla f = \lambda \nabla g.$$

- We use the condition $\nabla f = \lambda \nabla g$ to generate a system of equations, together with the constraint $g(x, y) = c$, that may be solved for x, y, and λ. Once we have all the solutions, we evaluate f at each of the (x, y) points to determine the extrema.

Exercises

1. The Cobb-Douglas production function is used in economics to model production levels based on labor and equipment. Suppose we have a specific Cobb-Douglas function of the form
$$f(x, y) = 50 x^{0.4} y^{0.6},$$
where x is the dollar amount spent on labor and y the dollar amount spent on equipment. Use the method of Lagrange multipliers to determine how much should be spent on labor and how much on equipment to maximize productivity if we have a total of $1.5 million dollars to invest in labor and equipment.

2. Use the method of Lagrange multipliers to find the point on the line $x - 2y = 5$ that is closest to the point $(1, 3)$. To do so, respond to the following prompts.

    (a) Write the function $f = f(x, y)$ that measures the *square* of the distance from (x, y) to $(1, 3)$. (The extrema of this function are the same as the extrema of the distance function, but $f(x, y)$ is simpler to work with.)

    (b) What is the constraint $g(x, y) = c$?

    (c) Write the equations resulting from $\nabla f = \lambda \nabla g$ and the constraint. Find all the points (x, y) satisfying these equations.

    (d) Test all the points you found to determine the extrema.

3. Apply the Method of Lagrange Multipliers solve each of the following constrained optimization problems.

    (a) Determine the absolute maximum and absolute minimum values of $f(x, y) = (x - 1)^2 + (y - 2)^2$ subject to the constraint that $x^2 + y^2 = 16$.

    (b) Determine the points on the sphere $x^2 + y^2 + z^2 = 4$ that are closest to and farthest from the point $(3, 1, -1)$. (As in the preceding exercise, you may find it simpler to work with the square of the distance formula, rather than the distance formula itself.)

(c) Find the absolute maximum and minimum of $f(x, y, z) = x^2 + y^2 + z^2$ subject to the constraint that $(x - 3)^2 + (y + 2)^2 + (z - 5)^2 \leq 16$. (Hint: here the constraint is a closed, bounded region. Use the boundary of that region for applying Lagrange Multipliers, but don't forget to also test any critical values of the function that lie in the interior of the region.)

4. There is a useful interpretation of the Lagrange multiplier λ. Assume that we want to optimize a function f with constraint $g(x, y) = c$. Recall that an optimal solution occurs at a point (x_0, y_0) where $\nabla f = \lambda \nabla g$. As the constraint changes, so does the point at which the optimal solution occurs. So we can think of the optimal point as a function of the parameter c, that is $x_0 = x_0(c)$ and $y_0 = y_0(c)$. The optimal value of f subject to the constraint can then be considered as a function of c defined by $f(x_0(c), y_0(c))$. The Chain Rule shows that

$$\frac{df}{dc} = \frac{\partial f}{\partial x_0} \frac{dx_0}{dc} + \frac{\partial f}{\partial y_0} \frac{dy_0}{dc}.$$

(a) Use the fact that $\nabla f = \lambda \nabla g$ at (x_0, y_0) to explain why

$$\frac{df}{dc} = \lambda \frac{dg}{dc}.$$

(b) Use the fact that $g(x, y) = c$ to show that

$$\frac{df}{dc} = \lambda.$$

Conclude that λ tells us the rate of change of the function f as the parameter c increases (or by approximately how much the optimal value of the function f will change if we increase the value of c by 1 unit).

(c) Suppose that $\lambda = 324$ at the point where the package described in Preview Activity 10.8 has its maximum volume. Explain in context what the value 324 tells us about the package.

Chapter 11

Multiple Integrals

11.1 Double Riemann Sums and Double Integrals over Rectangles

Motivating Questions

In this section, we strive to understand the ideas generated by the following important questions:

- What is a double Riemann sum?
- How is the double integral of a continuous function $f = f(x, y)$ defined?
- What are two things the double integral of a function can tell us?

Introduction

In single-variable calculus, recall that we approximated the area under the graph of a positive function f on an interval $[a, b]$ by adding areas of rectangles whose heights are determined by the curve. The general process involved subdividing the interval $[a, b]$ into smaller subintervals, constructing rectangles on each of these smaller intervals to approximate the region under the curve on that subinterval, then summing the areas of these rectangles to approximate the area under the curve. We will extend this process in this section to its three-dimensional analogs, double Riemann sums and double integrals over rectangles.

Preview Activity 11.1. In this activity we introduce the concept of a double Riemann sum.

(a) Review the concept of the Riemann sum from single-variable calculus. Then, explain how we define the definite integral $\int_a^b f(x)\, dx$ of a continuous function of a single variable x on an interval $[a, b]$. Include a sketch of a continuous function on an interval $[a, b]$ with appropriate labeling in order to illustrate your definition.

(b) In our upcoming study of integral calculus for multivariable functions, we will first extend the idea of the single-variable definite integral to functions of two variables over rectangular domains. To do so, we will need to understand how to partition a rectangle into subrectangles. Let R be rectangular domain $R = \{(x,y) : 0 \leq x \leq 6, 2 \leq y \leq 4\}$ (we can also represent this domain with the notation $[0,6] \times [2,4]$), as pictured in Figure 11.1.

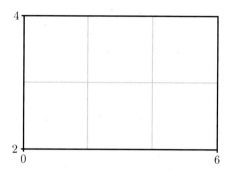

Figure 11.1: Rectangular domain R with subrectangles.

To form a partition of the full rectangular region, R, we will partition both intervals $[0, 6]$ and $[2, 4]$; in particular, we choose to partition the interval $[0, 6]$ into three uniformly sized subintervals and the interval $[2, 4]$ into two evenly sized subintervals as shown in Figure 11.1. In the following questions, we discuss how to identify the endpoints of each subinterval and the resulting subrectangles.

i. Let $0 = x_0 < x_1 < x_2 < x_3 = 6$ be the endpoints of the subintervals of $[0, 6]$ after partitioning. What is the length Δx of each subinterval $[x_{i-1}, x_i]$ for i from 1 to 3?

ii. Explicitly identify x_0, x_1, x_2, and x_3. On Figure 11.1 or your own version of the diagram, label these endpoints.

iii. Let $2 = y_0 < y_1 < y_2 = 4$ be the endpoints of the subintervals of $[2, 4]$ after partitioning. What is the length Δy of each subinterval $[y_{j-1}, y_j]$ for j from 1 to 2? Identify y_0, y_1, and y_2 and label these endpoints on Figure 11.1.

iv. Let R_{ij} denote the subrectangle $[x_{i-1}, x_i] \times [y_{j-1}, y_j]$. Appropriately label each subrectangle in your drawing of Figure 11.1. How does the total number of subrectangles depend on the partitions of the intervals $[0, 6]$ and $[2, 4]$?

v. What is area ΔA of each subrectangle?

Double Riemann Sums over Rectangles

For the definite integral in single-variable calculus, we considered a continuous function over a closed, bounded interval $[a, b]$. In multivariable calculus, we will eventually develop the idea of a definite integral over a closed, bounded region (such as the interior of a circle). We begin with a simpler situation by thinking only about rectangular domains, and will address more complicated domains in the following section.

Let $f = f(x, y)$ be a continuous function defined on a rectangular domain $R = \{(x, y) : a \leq x \leq b, c \leq y \leq d\}$. As we saw in Preview Activity 11.1, the domain is a rectangle R and we want to partition R into subrectangles. We do this by partitioning each of the intervals $[a, b]$ and $[c, d]$ into subintervals and using those subintervals to create a partition of R into subrectangles. In the first activity, we address the quantities and notations we will use in order to define double Riemann sums and double integrals.

Activity 11.1.

Let $f(x, y) = 100 - x^2 - y^2$ be defined on the rectangular domain $R = [a, b] \times [c, d]$. Partition the interval $[a, b]$ into four uniformly sized subintervals and the interval $[c, d]$ into three evenly sized subintervals as shown in Figure 11.2. As we did in Preview Activity 11.1, we will need a method for identifying the endpoints of each subinterval and the resulting subrectangles.

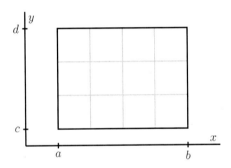

Figure 11.2: Rectangular domain with subrectangles.

(a) Let $a = x_0 < x_1 < x_2 < x_3 < x_4 = b$ be the endpoints of the subintervals of $[a, b]$ after partitioning. Label these endpoints in Figure 11.2.

(b) What is the length Δx of each subinterval $[x_{i-1}, x_i]$? Your answer should be in terms of a and b.

(c) Let $c = y_0 < y_1 < y_2 < y_3 = d$ be the endpoints of the subintervals of $[c, d]$ after partitioning. Label these endpoints in Figure 11.2.

(d) What is the length Δy of each subinterval $[y_{j-1}, y_j]$? Your answer should be in terms of c and d.

(e) The partitions of the intervals $[a, b]$ and $[c, d]$ partition the rectangle R into subrectangles. How many subrectangles are there?

(f) Let R_{ij} denote the subrectangle $[x_{i-1}, x_i] \times [y_{j-1}, y_j]$. Label each subrectangle in Figure 11.2.

(g) What is area ΔA of each subrectangle?

(h) Now let $[a,b] = [0,8]$ and $[c,d] = [2,6]$. Let (x_{11}^*, y_{11}^*) be the point in the upper right corner of the subrectangle R_{11}. Identify and correctly label this point in Figure 11.2. Calculate the product

$$f(x_{11}^*, y_{11}^*)\Delta A.$$

Explain, geometrically, what this product represents.

(i) For each i and j, choose a point (x_{ij}^*, y_{ij}^*) in the subrectangle $R_{i,j}$. Identify and correctly label these points in Figure 11.2. Explain what the product

$$f(x_{ij}^*, y_{ij}^*)\Delta A$$

represents.

(j) If we were to add all the values $f(x_{ij}^*, y_{ij}^*)\Delta A$ for each i and j, what does the resulting number approximate about the surface defined by f on the domain R? (You don't actually need to add these values.)

(k) Write a double sum using summation notation that expresses the arbitrary sum from part (j).

◁

Double Riemann Sums and Double Integrals

Now we use the process from the most recent activity to formally define double Riemann sums and double integrals.

> **Definition 11.1.** Let f be a continuous function on a rectangle $R = \{(x,y) : a \leq x \leq b, c \leq y \leq d\}$. A **double Riemann sum for f over R** is created as follows.
>
> - Partition the interval $[a,b]$ into m subintervals of equal length $\Delta x = \frac{b-a}{m}$. Let $x_0, x_1, \ldots, x_m$ be the endpoints of these subintervals, where $a = x_0 < x_1 < x_2 < \cdots < x_m = b$.
>
> - Partition the interval $[c,d]$ into n subintervals of equal length $\Delta y = \frac{d-c}{n}$. Let $y_0, y_1, \ldots, y_n$ be the endpoints of these subintervals, where $c = y_0 < y_1 < y_2 < \cdots < y_n = d$.
>
> - These two partitions create a partition of the rectangle R into mn subrectangles R_{ij} with opposite vertices (x_{i-1}, y_{j-1}) and (x_i, y_j) for i between 1 and m and j between 1 and n. These rectangles all have equal area $\Delta A = \Delta x \cdot \Delta y$.
>
> - Choose a point (x_{ij}^*, y_{ij}^*) in each rectangle R_{ij}. Then, a double Riemann sum for f over R is given by
>
> $$\sum_{j=1}^{n} \sum_{i=1}^{m} f(x_{ij}^*, y_{ij}^*) \cdot \Delta A.$$

11.1. DOUBLE RIEMANN SUMS AND DOUBLE INTEGRALS OVER RECTANGLES

If $f(x,y) \geq 0$ on the rectangle R, we may ask to find the volume of the solid bounded above by f over R, as illustrated on the left of Figure 11.3. This volume is approximated by a Riemann sum, which sums the volumes of the rectangular boxes shown on the right of Figure 11.3.

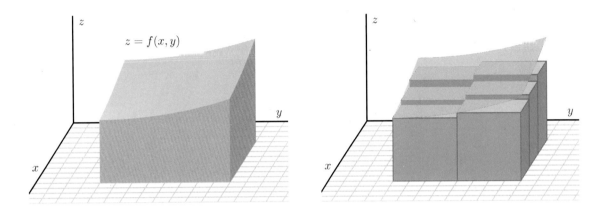

Figure 11.3: The volume under a graph approximated by a Riemann Sum

As we let the number of subrectangles increase without bound (in other words, as both m and n in a double Riemann sum go to infinity), as illustrated in Figure 11.4, the sum of the volumes of the rectangular boxes approaches the volume of the solid bounded above by f over R. The value of this limit, provided it exists, is the double integral.

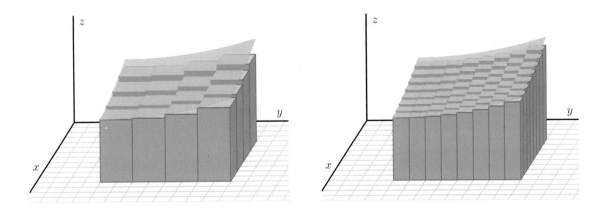

Figure 11.4: Finding better approximations by using smaller subrectangles.

Definition 11.2. Let R be a rectangular region in the x-y plane and f a continuous function over R. With terms defined as in a double Riemann sum, the **double integral of f over R** is

$$\iint_R f(x,y)\, dA = \lim_{m,n \to \infty} \sum_{j=1}^{n} \sum_{i=1}^{m} f(x_{ij}^*, y_{ij}^*) \cdot \Delta A.$$

Interpretation of Double Riemann Sums and Double integrals.

At the moment, there are two ways we can interpret the value of the double integral.

- Suppose that $f(x,y)$ assumes both positive and negatives values on the rectangle R, as shown on the left of Figure 11.5. When constructing a Riemann sum, for each i and j, the product $f(x_{ij}^*, y_{ij}^*) \cdot \Delta A$ can be interpreted as a "signed" volume of a box with base area ΔA and "signed" height $f(x_{ij}^*, y_{ij}^*)$. Since f can have negative values, this "height" could be negative. The sum

$$\sum_{j=1}^{n} \sum_{i=1}^{m} f(x_{ij}^*, y_{ij}^*) \cdot \Delta A$$

can then be interpreted as a sum of "signed" volumes of boxes, with a negative sign attached to those boxes whose heights are below the xy-plane. We can then realize the double integral

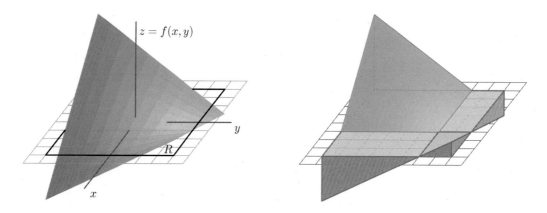

Figure 11.5: The integral measures signed volume.

$\iint_R f(x,y)\, dA$ as a difference in volumes: $\iint_R f(x,y)\, dA$ tells us the volume of the solids the graph of f bounds above the xy-plane over the rectangle R minus the volume of the solids the graph of f bounds below the xy-plane under the rectangle R. This is shown on the right of Figure 11.5.

- The average of the finitely many mn values $f(x^*_{ij}, y^*_{ij})$ that we take in a double Riemann sum is given by

$$\text{Avg}_{mn} = \frac{1}{mn} \sum_{j=1}^{n} \sum_{i=1}^{m} f(x^*_{ij}, y^*_{ij}).$$

If we take the limit as m and n go to infinity, we obtain what we define as the average value of f over the region R, which is connected to the value of the double integral. First, to view Avg_{mn} as a double Riemann sum, note that

$$\Delta x = \frac{b-a}{m} \quad \text{and} \quad \Delta y = \frac{d-c}{n}.$$

Thus,

$$\frac{1}{mn} = \frac{\Delta x \cdot \Delta y}{(b-a)(d-c)} = \frac{\Delta A}{A(R)},$$

where $A(R)$ denotes the area of the rectangle R. Then, the average value of the function f over R, $f_{\text{AVG}(R)}$, is given by

$$f_{\text{AVG}(R)} = \lim_{m,n \to \infty} \frac{1}{mn} \sum_{j=1}^{n} \sum_{i=1}^{m} f(x^*_{ij}, y^*_{ij})$$

$$= \lim_{m,n \to \infty} \frac{1}{A(R)} \sum_{j=1}^{n} \sum_{i=1}^{m} f(x^*_{ij}, y^*_{ij}) \cdot \Delta A$$

$$= \frac{1}{A(R)} \iint_R f(x,y) \, dA.$$

Therefore, the double integral of f over R divided by the area of R gives us the average value of the function f on R. Finally, if $f(x,y) \geq 0$ on R, we can interpret this average value of f on R as the height of the box with base R that has the same volume as the volume of the surface defined by f over R.

Activity 11.2.

Let $f(x,y) = x + 2y$ and let $R = [0,2] \times [1,3]$.

(a) Draw a picture of R. Partition $[0,2]$ into 2 subintervals of equal length and the interval $[1,3]$ into two subintervals of equal length. Draw these partitions on your picture of R and label the resulting subrectangles using the labeling scheme we established in the definition of a double Riemann sum.

(b) For each i and j, let (x^*_{ij}, y^*_{ij}) be the midpoint of the rectangle R_{ij}. Identify the coordinates of each (x^*_{ij}, y^*_{ij}). Draw these points on your picture of R.

(c) Calculate the Riemann sum
$$\sum_{j=1}^{n} \sum_{i=1}^{m} f(x^*_{ij}, y^*_{ij}) \cdot \Delta A$$

using the partitions we have described. If we let (x^*_{ij}, y^*_{ij}) be the midpoint of the rectangle R_{ij} for each i and j, then the resulting Riemann sum is called a *midpoint sum*.

(d) Give two interpretations for the meaning of the sum you just calculated.

◁

Activity 11.3.

Let $f(x,y) = \sqrt{4-y^2}$ on the rectangular domain $R = [1,7] \times [-2,2]$. Partition $[1,7]$ into 3 equal length subintervals and $[-2,2]$ into 2 equal length subintervals. A table of values of f at some points in R is given in Table 11.1, and a graph of f with the indicated partitions is shown in Figure 11.6.

	-2	-1	0	1	2
1	0	$\sqrt{3}$	2	$\sqrt{3}$	0
2	0	$\sqrt{3}$	2	$\sqrt{3}$	0
3	0	$\sqrt{3}$	2	$\sqrt{3}$	0
4	0	$\sqrt{3}$	2	$\sqrt{3}$	0
5	0	$\sqrt{3}$	2	$\sqrt{3}$	0
6	0	$\sqrt{3}$	2	$\sqrt{3}$	0
7	0	$\sqrt{3}$	2	$\sqrt{3}$	0

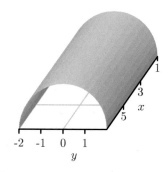

Table 11.1: Table of values of $f(x,y) = \sqrt{4-y^2}$.

Figure 11.6: Graph of $f(x,y) = \sqrt{4-y^2}$ on R.

(a) Outline the partition of R into subrectangles on the table of values in Table 11.1.

(b) Calculate the double Riemann sum using the given partition of R and the values of f in the upper right corner of each subrectangle.

(c) Use geometry to calculate the exact value of $\iint_R f(x,y)\, dA$ and compare it to your approximation. How could we obtain a better approximation?

◁

We conclude this section with a list of properties of double integrals. Since similar properties are satisfied by single-variable integrals and the arguments for double integrals are essentially the

11.1. DOUBLE RIEMANN SUMS AND DOUBLE INTEGRALS OVER RECTANGLES

same, we omit their justification.

> **Properties of Double Integrals.** Let f and g be continuous functions on a rectangle $R = \{(x,y) : a \leq x \leq b, c \leq y \leq d\}$, and let k be a constant. Then
>
> 1. $\iint_R (f(x,y) + g(x,y))\, dA = \iint_R f(x,y)\, dA + \iint_R g(x,y)\, dA$.
> 2. $\iint_R k f(x,y)\, dA = k \iint_R f(x,y)\, dA$.
> 3. If $f(x,y) \geq g(x,y)$ on R, then $\iint_R f(x,y)\, dA \geq \iint_R g(x,y)\, dA$.

Summary

- Let f be a continuous function on a rectangle $R = \{(x,y) : a \leq x \leq b, c \leq y \leq d\}$. The double Riemann sum for f over R is created as follows.

    - Partition the interval $[a,b]$ into m subintervals of equal length $\Delta x = \frac{b-a}{m}$. Let $x_0, x_1, \ldots, x_m$ be the endpoints of these subintervals, where $a = x_0 < x_1 < x_2 < \cdots < x_m = b$.

    - Partition the interval $[c,d]$ into n subintervals of equal length $\Delta y = \frac{d-c}{n}$. Let $y_0, y_1, \ldots, y_n$ be the endpoints of these subintervals, where $c = y_0 < y_1 < y_2 < \cdots < y_n = d$.

    - These two partitions create a partition of the rectangle R into mn subrectangles R_{ij} with opposite vertices (x_{i-1}, y_{j-1}) and (x_i, y_j) for i between 1 and m and j between 1 and n. These rectangles all have equal area $\Delta A = \Delta x \cdot \Delta y$.

    - Choose a point (x_{ij}^*, y_{ij}^*) in each rectangle R_{ij}. Then a double Riemann sum for f over R is given by
    $$\sum_{j=1}^{n} \sum_{i=1}^{m} f(x_{ij}^*, y_{ij}^*) \cdot \Delta A.$$

- With terms defined as in the Double Riemann Sum, the double integral of f over R is
$$\iint_R f(x,y)\, dA = \lim_{m,n \to \infty} \sum_{j=1}^{n} \sum_{i=1}^{m} f(x_{ij}^*, y_{ij}^*) \cdot \Delta A.$$

- Two interpretations of the double integral $\iint_R f(x,y)\, dA$ are:

    - The volume of the solids the graph of f bounds above the xy-plane over the rectangle R minus the volume of the solids the graph of f bounds below the xy-plane under the rectangle R;

    - Dividing the double integral of f over R by the area of R gives us the average value of the function f on R. If $f(x,y) \geq 0$ on R, we can interpret this average value of f on R as the height of the box with base R that has the same volume as the volume of the surface defined by f over R.

Exercises

1. The temperature at any point on a metal plate in the xy plane is given by $T(x, y) = 100 - 4x^2 - y^2$, where x and y are measured in inches and T in degrees Celsius. Consider the portion of the plate that lies on the rectangular region $R = [1, 5] \times [3, 6]$.

   (a) Estimate the value of $\iint_R T(x, y)\, dA$ by using a double Riemann sum with two subintervals in each direction and choosing (x_i^*, y_j^*) to be the point that lies in the upper right corner of each subrectangle.

   (b) Determine the area of the rectangle R.

   (c) Estimate the average temperature, $T_{\text{AVG}(R)}$, over the region R.

   (d) Do you think your estimate in (c) is an over- or under-estimate of the true temperature? Why?

2. The wind chill, as frequently reported, is a measure of how cold it feels outside when the wind is blowing. In Table 11.2, the wind chill $w = w(v, T)$, measured in degrees Fahrenheit, is a function of the wind speed v, measured in miles per hour, and the ambient air temperature T, also measured in degrees Fahrenheit. Approximate the average wind chill on the rectangle $[5, 35] \times [-20, 20]$ using 3 subintervals in the v direction, 4 subintervals in the T direction, and the point in the lower left corner in each subrectangle.

$v \backslash T$	-20	-15	-10	-5	0	5	10	15	20
5	-34	-28	-22	-16	-11	-5	1	7	13
10	-41	-35	-28	-22	-16	-10	-4	3	9
15	-45	-39	-32	-26	-19	-13	-7	0	6
20	-48	-42	-35	-29	-22	-15	-9	-2	4
25	-51	-44	-37	-31	-24	-17	-11	-4	3
30	-53	-46	-39	-33	-26	-19	-12	-5	1
35	-55	-48	-41	-34	-27	-21	-14	-7	0

Table 11.2: Wind chill as a function of wind speed and temperature.

3. Consider the box with a sloped top that is given by the following description: the base is the rectangle $R = [0, 4] \times [0, 3]$, while the top is given by the plane $z = p(x, y) = 20 - 2x - 3y$.

   (a) Estimate the value of $\iint_R p(x, y)\, dA$ by using a double Riemann sum with four subintervals in the x direction and three subintervals in the y direction, and choosing (x_i^*, y_j^*) to be the point that is the midpoint of each subrectangle.

   (b) What important quantity does your double Riemann sum in (a) estimate?

(c) Suppose it can be determined that $\iint_R p(x,y)\,dA = 138$. What is the exact average value of p over R?

(d) If you wanted to build a rectangular box (with the same base) that has the same volume as the box with the sloped top described here, how tall would the rectangular box have to be?

11.2 Iterated Integrals

> **Motivating Questions**
>
> *In this section, we strive to understand the ideas generated by the following important questions:*
>
> - How do we evaluate a double integral over a rectangle as an iterated integral, and why does this process work?

Introduction

Recall that we defined the double integral of a continuous function $f = f(x,y)$ over a rectangle $R = [a,b] \times [c,d]$ as

$$\iint_R f(x,y)\, dA = \lim_{m,n \to \infty} \sum_{j=1}^{n} \sum_{i=1}^{m} f\left(x_{ij}^*, y_{ij}^*\right) \cdot \Delta A,$$

where the different variables and notation are as described in Section 11.1. Thus $\iint_R f(x,y)\, dA$ is a limit of double Riemann sums, but while this definition tells us exactly what a double integral is, it is not very helpful for determining the value of a double integral. Fortunately, there is a way to view a double integral as an *iterated integral*, which will make computations feasible in many cases.

The viewpoint of an iterated integral is closely connected to an important idea from single-variable calculus. When we studied solids of revolution, such as the one shown in Figure 11.7, we saw that in some circumstances we could slice the solid perpendicular to an axis and have each slice be approximately a circular disk. From there, we were able to find the volume of each disk, and then use an integral to add the volumes of the slices. In what follows, we are able to use single integrals to generalize this approach to handle even more general geometric shapes.

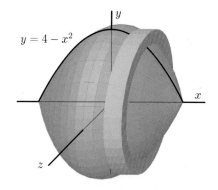

Figure 11.7: A solid of revolution.

11.2. ITERATED INTEGRALS

Preview Activity 11.2. Let $f(x,y) = 25 - x^2 - y^2$ on the rectangular domain $R = [-3, 3] \times [-4, 4]$.

As with partial derivatives, we may treat one of the variables in f as constant and think of the resulting function as a function of a single variable. Now we investigate what happens if we integrate instead of differentiate.

(a) Choose a fixed value of x in the interior of $[-3, 3]$. Let

$$A(x) = \int_{-4}^{4} f(x, y) \, dy.$$

What is the geometric meaning of the value of $A(x)$ relative to the surface defined by f. (Hint: Think about the trace determined by the fixed value of x, and consider how $A(x)$ is related to Figure 11.8.)

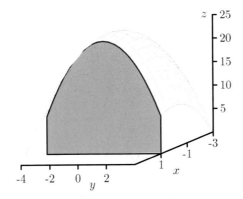

Figure 11.8: A cross section with fixed x.

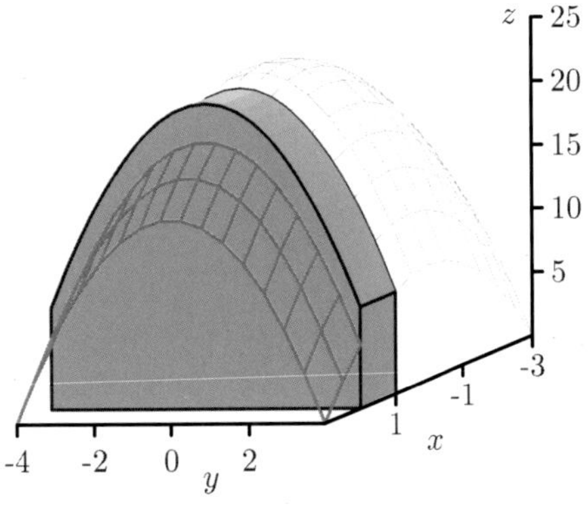

Figure 11.9: A cross section with fixed x and Δx.

(b) For a fixed value of x, say x_i^*, what is the geometric meaning of $A(x_i^*) \Delta x$? (Hint: Consider how $A(x_i^*) \Delta x$ is related to Figure 11.9.)

(c) Since f is continuous on R, we can define the function $A = A(x)$ at every value of x in $[-3, 3]$. Now think about subdividing the x-interval $[-3, 3]$ into m subintervals, and choosing a value x_i^* in each of those subintervals. What will be the meaning of the sum

$$\sum_{i=1}^{m} A(x_i^*) \Delta x?$$

(d) Explain why $\int_{-3}^{3} A(x) \, dx$ will determine the exact value of the volume under the surface $z = f(x, y)$ over the rectangle R.

Iterated Integrals

The ideas that we explored in Preview Activity 11.2 work more generally and lead to the idea of an iterated integral. Let f be a continuous function on a rectangular domain $R = [a, b] \times [c, d]$, and let

$$A(x) = \int_c^d f(x, y)\, dy.$$

The function $A = A(x)$ determines the value of the cross sectional area[1] in the y direction for the fixed value of x of the solid bounded between the surface defined by f and the xy-plane.

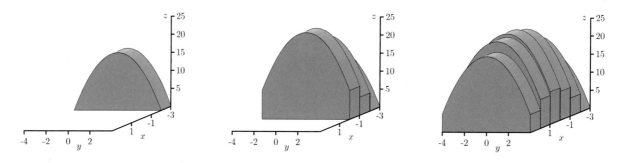

Figure 11.10: Summing cross section slices.

The value of this cross sectional area is determined by the input x in A. Since A is a function of x, it follows that we can integrate A with respect to x. In doing so, we use a partition of $[a, b]$ and make an approximation to the integral given by

$$\int_a^b A(x)\, dx \approx \sum_{i=1}^m A(x_i^*) \Delta x,$$

where x_i^* is any number in the subinterval $[x_{i-1}, x_i]$. Each term $A(x_i^*)\Delta x$ in the sum represents an approximation of a fixed cross sectional slice of the surface in the y direction with a fixed width of Δx as illustrated in Figure 11.9. We add the signed volumes of these slices as shown in the frames in Figure 11.10 to obtain an approximation of the total signed volume.

As we let the number of subintervals in the x direction approach infinity, we can see that the Riemann sum $\sum_{i=1}^m A(x_i^*) \Delta x$ approaches a limit and that limit is the sum of signed volumes bounded by the function f on R. Therefore, since $A(x)$ is itself determined by an integral, we have

$$\iint_R f(x, y)\, dA = \lim_{m \to \infty} \sum_{i=1}^m A(x_i^*) \Delta x = \int_a^b A(x)\, dx = \int_a^b \left(\int_c^d f(x, y)\, dy \right) dx.$$

---

[1] By area we mean "signed" area.

11.2. ITERATED INTEGRALS

Hence, we can compute the double integral of f over R by first integrating f with respect to y on $[c, d]$, then integrating the resulting function of x with respect to x on $[a, b]$. The nested integral

$$\int_a^b \left(\int_c^d f(x, y) \, dy \right) dx = \int_a^b \int_c^d f(x, y) \, dy \, dx$$

is called an *iterated integral*, and we see that each double integral may be represented by two single integrals.

We made a choice to integrate first with respect to y. The same argument shows that we can also find the double integral as an iterated integral integrating with respect to x first, or

$$\iint_R f(x, y) \, dA = \int_c^d \left(\int_a^b f(x, y) \, dx \right) dy = \int_c^d \int_a^b f(x, y) \, dx \, dy.$$

The fact that integrating in either order results in the same value is known as Fubini's Theorem.

Fubini's Theorem. If $f = f(x, y)$ is a continuous function on a rectangle $R = [a, b] \times [c, d]$, then

$$\iint_R f(x, y) \, dA = \int_c^d \int_a^b f(x, y) \, dx \, dy = \int_a^b \int_c^d f(x, y) \, dy \, dx.$$

Fubini's theorem enables us to evaluate iterated integrals without resorting to the limit definition. Instead, working with one integral at a time, we can use the Fundamental Theorem of Calculus from single-variable calculus to find the exact value of each integral, starting with the inner integral.

Activity 11.4.

Let $f(x, y) = 25 - x^2 - y^2$ on the rectangular domain $R = [-3, 3] \times [-4, 4]$.

(a) Viewing x as a fixed constant, use the Fundamental Theorem of Calculus to evaluate the integral

$$A(x) = \int_{-4}^4 f(x, y) \, dy.$$

Note that you will be integrating with respect to y, and holding x constant. Your result should be a function of x only.

(b) Next, use your result from (a) along with the Fundamental Theorem of Calculus to determine the value of $\int_{-3}^3 A(x) \, dx$.

(c) What is the value of $\iint_R f(x, y) \, dA$? What are two different ways we may interpret the meaning of this value?

◁

Activity 11.5.

Let $f(x,y) = x + y^2$ on the rectangle $R = [0,2] \times [0,3]$.

(a) Evaluate $\iint_R f(x,y)\,dA$ using an iterated integral. Choose an order for integration by deciding whether you want to integrate first with respect to x or y.

(b) Evaluate $\iint_R f(x,y)\,dA$ using the iterated integral whose order of integration is the opposite of the order you chose in (a).

◁

Summary

- We can evaluate the double integral $\iint_R f(x,y)\,dA$ over a rectangle $R = [a,b] \times [c,d]$ as an iterated integral in one of two ways:

  - $\int_a^b \left(\int_c^d f(x,y)\,dy \right) dx$, or

  - $\int_c^d \left(\int_a^b f(x,y)\,dx \right) dy$.

  This process works because each inner integral represents a cross-sectional (signed) area and the outer integral then sums all of the cross-sectional (signed) areas. Fubini's Theorem guarantees that the resulting value is the same, regardless of the order in which we integrate.

Exercises

1. Evaluate each of the following double or iterated integrals exactly.

   (a) $\int_1^3 \left(\int_2^5 xy\,dy \right) dx$

   (b) $\int_0^{\pi/4} \left(\int_0^{\pi/3} \sin(x)\cos(y)\,dx \right) dy$

   (c) $\int_0^1 \left(\int_0^1 e^{-2x-3y}\,dy \right) dx$

   (d) $\iint_R \sqrt{2x + 5y}\,dA$, where $R = [0,2] \times [0,3]$.

2. The temperature at any point on a metal plate in the xy plane is given by $T(x,y) = 100 - 4x^2 - y^2$, where x and y are measured in inches and T in degrees Celsius. Consider the portion of the plate that lies on the rectangular region $R = [1,5] \times [3,6]$.

   (a) Write an iterated integral whose value represents the volume under the surface T over the rectangle R.

11.2. ITERATED INTEGRALS

(b) Evaluate the iterated integral you determined in (a).

(c) Find the area of the rectangle, R.

(d) Determine the exact average temperature, $T_{\text{AVG}(R)}$, over the region R.

3. Consider the box with a sloped top that is given by the following description: the base is the rectangle $R = [1, 4] \times [2, 5]$, while the top is given by the plane $z = p(x, y) = 30 - x - 2y$.

    (a) Write an iterated integral whose value represents the volume under p over the rectangle R.

    (b) Evaluate the iterated integral you determined in (a).

    (c) What is the exact average value of p over R?

    (d) If you wanted to build a rectangular box (with an identical base) that has the same volume as the box with the sloped top described here, how tall would the rectangular box have to be?

11.3 Double Integrals over General Regions

Motivating Questions

In this section, we strive to understand the ideas generated by the following important questions:

- How do we define a double integral over a non-rectangular region?
- What general form does an iterated integral over a non-rectangular region have?

Introduction

Recall that we defined the double integral of a continuous function $f = f(x, y)$ over a rectangle $R = [a, b] \times [c, d]$ as

$$\iint_R f(x, y)\, dA = \lim_{m,n \to \infty} \sum_{j=1}^{n} \sum_{i=1}^{m} f(x_{ij}^*, y_{ij}^*) \cdot \Delta A,$$

where the notation is as described in Section 11.1. Furthermore, we have seen that we can evaluate a double integral $\iint_R f(x, y)\, dA$ over R as an iterated integral of either of the forms

$$\int_a^b \int_c^d f(x, y)\, dy\, dx \quad \text{or} \quad \int_c^d \int_a^b f(x, y)\, dx\, dy.$$

It is natural to wonder how we might define and evaluate a double integral over a non-rectangular region; we explore one such example in the following preview activity.

Preview Activity 11.3. A tetrahedron is a three-dimensional figure with four faces, each of which is a triangle. A picture of the tetrahedron T with vertices $(0, 0, 0)$, $(1, 0, 0)$, $(0, 1, 0)$, and $(0, 0, 1)$ is shown in Figure 11.11. If we place one vertex at the origin and let vectors **a**, **b**, and **c** be determined by the edges of the tetrahedron that have one end at the origin, then a formula that tells us the volume V of the tetrahedron is

$$V = \frac{1}{6} |\mathbf{a} \cdot (\mathbf{b} \times \mathbf{c})|. \tag{11.1}$$

(a) Use the formula (11.1) to find the volume of the tetrahedron T.

(b) Instead of memorizing or looking up the formula for the volume of a tetrahedron, we can use a double integral to calculate the volume of the tetrahedron T. To see how, notice that the top face of the tetrahedron T is the plane whose equation is

$$z = 1 - (x + y).$$

11.3. DOUBLE INTEGRALS OVER GENERAL REGIONS

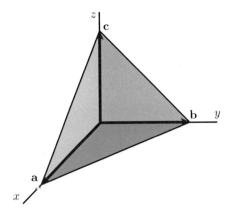

Figure 11.11: The tetrahedron T.

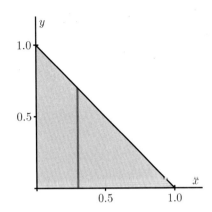

Figure 11.12: Projecting T onto the xy-plane.

Provided that we can use an iterated integral on a non-rectangular region, the volume of the tetrahedron will be given by an iterated integral of the form

$$\int_{x=?}^{x=?} \int_{y=?}^{y=?} 1 - (x+y)\, dy\, dx.$$

The issue that is new here is how we find the limits on the integrals; note that the outer integral's limits are in x, while the inner ones are in y, since we have chosen $dA = dy\, dx$. To see the domain over which we need to integrate, think of standing way above the tetrahedron looking straight down on it, which means we are projecting the entire tetrahedron onto the xy-plane. The resulting domain is the triangular region shown in Figure 11.12.

Explain why we can represent the triangular region with the inequalities

$$0 \le y \le 1 - x \quad \text{and} \quad 0 \le x \le 1.$$

(Hint: Consider the cross sectional slice shown in Figure 11.12.)

(c) Explain why it makes sense to now write the volume integral in the form

$$\int_{x=?}^{x=?} \int_{y=?}^{y=?} 1 - (x+y)\, dy\, dx = \int_{x=0}^{x=1} \int_{y=0}^{y=1-x} 1 - (x+y)\, dy\, dx.$$

(d) Use the Fundamental Theorem of Calculus to evaluate the iterated integral

$$\int_{x=0}^{x=1} \int_{y=0}^{y=1-x} 1 - (x+y)\, dy\, dx$$

and compare to your result from part (a). (As with iterated integrals over rectangular regions, start with the inner integral.)

Double Integrals over General Regions

So far, we have learned that a double integral over a rectangular region may be interpreted in one of two ways:

- $\iint_R f(x, y)\, dA$ tells us the volume of the solids the graph of f bounds above the xy-plane over the rectangle R minus the volume of the solids the graph of f bounds below the xy-plane under the rectangle R;

- $\frac{1}{A(R)} \iint_R f(x, y)\, dA$, where $A(R)$ is the area of R tells us the average value of the function f on R. If $f(x, y) \geq 0$ on R, we can interpret this average value of f on R as the height of the box with base R that has the same volume as the volume of the surface defined by f over R.

As we saw in Preview Activity 11.1, a function $f = f(x, y)$ may be considered over regions other than rectangular ones, and thus we want to understand how to set up and evaluate double integrals over non-rectangular regions. Note that if we can, then the two interpretations of the double integral noted above will naturally extend to solid regions with non-rectangular bases.

So, suppose f is a continuous function on a closed, bounded domain D. For example, consider D as the circular domain shown in Figure 11.13. We can enclose D in a rectangular domain R as

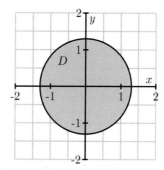

Figure 11.13: A non-rectangular domain.

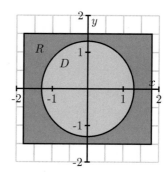

Figure 11.14: Enclosing this domain in a rectangle.

shown in Figure 11.14 and extend the function f to be defined over R in order to be able to use the definition of the double integral over a rectangle. We extend f in such a way that its values at the points in R that are not in D contribute 0 to the value of the integral. In other words, define a

11.3. DOUBLE INTEGRALS OVER GENERAL REGIONS

function $F = F(x, y)$ on R as

$$F(x, y) = \begin{cases} f(x, y), & \text{if } (x, y) \in D, \\ 0, & \text{if } (x, y) \notin D \end{cases}.$$

We then say that the double integral of f over D is the same as the double integral of F over R, and thus

$$\iint_D f(x, y)\, dA = \iint_R F(x, y)\, dA.$$

In practice, we just ignore everything that is in R but not in D, since these regions contribute 0 to the value of the integral.

Just as with double integrals over rectangles, a double integral over a domain D can be evaluated as an iterated integral. If the region D can be described by the inequalities $g_1(x) \leq y \leq g_2(x)$ and $a \leq x \leq b$, where $g_1 = g_1(x)$ and $g_2 = g_2(x)$ are functions of only x, then

$$\iint_D f(x, y)\, dA = \int_{x=a}^{x=b} \int_{y=g_1(x)}^{y=g_2(x)} f(x, y)\, dy\, dx.$$

Alternatively, if the region D is described by the inequalities $h_1(y) \leq x \leq h_2(y)$ and $c \leq y \leq d$, where $h_1 = h_1(y)$ and $h_2 = h_2(y)$ are functions of only y, we have

$$\iint_D f(x, y)\, dA = \int_{y=c}^{y=d} \int_{x=h_1(y)}^{x=h_2(y)} f(x, y)\, dx\, dy.$$

The structure of an iterated integral is of particular note:

In an iterated double integral:

- the limits on the outer integral must be constants;

- the limits on the inner integral must be constants or in terms of only the remaining variable – that is, if the inner integral is with respect to y, then its limits may only involve x and constants, and vice versa.

We next consider a detailed example.

Example 11.1. Let $f(x, y) = x^2 y$ be defined on the triangle D with vertices $(0, 0)$, $(2, 0)$, and $(2, 3)$ as shown in Figure 11.15. To evaluate $\iint_D f(x, y)\, dA$, we must first describe the region D in terms of the variables x and y. We take two approaches.

Approach 1: Integrate first with respect to y. In this case we choose to evaluate the double integral as an iterated integral in the form

$$\iint_D x^2 y\, dA = \int_{x=a}^{x=b} \int_{y=g_1(x)}^{y=g_2(x)} x^2 y\, dy\, dx,$$

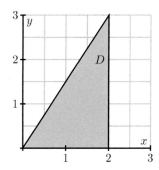
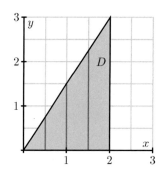
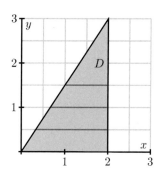

Figure 11.15: A triangular domain.

Figure 11.16: Slices in the y direction.

Figure 11.17: Slices in the x direction.

and therefore we need to describe D in terms of inequalities

$$g_1(x) \leq y \leq g_2(x) \quad \text{and} \quad a \leq x \leq b.$$

Since we are integrating with respect to y first, the iterated integral has the form

$$\iint_D x^2 y \, dA = \int_{x=a}^{x=b} A(x) \, dx,$$

where $A(x)$ is a cross sectional area in the y direction. So we are slicing the domain perpendicular to the x-axis and want to understand what a cross sectional area of the overall solid will look like. Several slices of the domain are shown in Figure 11.16. On a slice with fixed x value, the y values are bounded below by 0 and above by the y coordinate on the hypotenuse of the right triangle. Thus, $g_1(x) = 0$; to find $y = g_2(x)$, we need to write the hypotenuse as a function of x. The hypotenuse connects the points (0,0) and (2,3) and hence has equation $y = \frac{3}{2}x$. This gives the upper bound on y as $g_2(x) = \frac{3}{2}x$. The leftmost vertical cross section is at $x = 0$ and the rightmost one is at $x = 2$, so we have $a = 0$ and $b = 2$. Therefore,

$$\iint_D x^2 y \, dA = \int_{x=0}^{x=2} \int_{y=0}^{y=\frac{3}{2}x} x^2 y \, dy \, dx.$$

We evaluate the iterated integral by applying the Fundamental Theorem of Calculus first to

11.3. DOUBLE INTEGRALS OVER GENERAL REGIONS

the inner integral, and then to the outer one, and find that

$$\int_{x=0}^{x=2} \int_{y=0}^{y=\frac{3}{2}x} x^2 y\, dy\, dx = \int_{x=0}^{x=2} \left[x^2 \cdot \frac{y^2}{2}\right]\bigg|_{y=0}^{y=\frac{3}{2}x} dx$$

$$= \int_{x=0}^{x=2} \frac{9}{8} x^4\, dx$$

$$= \frac{9}{8} \frac{x^5}{5}\bigg|_{x=0}^{x=2}$$

$$= \left(\frac{9}{8}\right)\left(\frac{32}{5}\right)$$

$$= \frac{36}{5}.$$

Approach 2: Integrate first with respect to x. In this case, we choose to evaluate the double integral as an iterated integral in the form

$$\iint_D x^2 y\, dA = \int_{y=c}^{y=d} \int_{x=h_1(y)}^{x=h_2(y)} x^2 y\, dx\, dy$$

and thus need to describe D in terms of inequalities

$$h_1(y) \le x \le h_2(y) \quad \text{and} \quad c \le y \le d.$$

Since we are integrating with respect to x first, the iterated integral has the form

$$\iint_D x^2 y\, dA = \int_c^d A(y)\, dy,$$

where $A(y)$ is a cross sectional area of the solid in the x direction. Several slices of the domain – perpendicular to the y-axis – are shown in Figure 11.17.

On a slice with fixed y value, the x values are bounded below by the x coordinate on the hypotenuse of the right triangle and above by 2. So $h_2(y) = 2$; to find $h_1(y)$, we need to write the hypotenuse as a function of y. Solving the earlier equation we have for the hypotenuse ($y = \frac{3}{2}x$) for x gives us $x = \frac{2}{3}y$. This makes $h_1(y) = \frac{2}{3}y$. The lowest horizontal cross section is at $y = 0$ and the uppermost one is at $y = 3$, so we have $c = 0$ and $d = 3$. Therefore,

$$\iint_D x^2 y\, dA = \int_{y=0}^{y=3} \int_{x=(2/3)y}^{x=2} x^2 y\, dx\, dy.$$

We evaluate the resulting iterated integral as before by twice applying the Fundamental

Theorem of Calculus, and find that

$$\int_{y=0}^{y=3} \int_{x=\frac{2}{3}y}^{2} x^2 y \, dx \, dy = \int_{y=0}^{y=3} \left[\frac{x^3}{3}\right]\Big|_{x=\frac{2}{3}y}^{x=2} y \, dx$$

$$= \int_{y=0}^{y=3} \left[\frac{8}{3}y - \frac{8}{81}y^4\right] dy$$

$$= \left[\frac{8}{3}\frac{y^2}{2} - \frac{8}{81}\frac{y^5}{5}\right]\Big|_{y=0}^{y=3}$$

$$= \left(\frac{8}{3}\right)\left(\frac{9}{2}\right) - \left(\frac{8}{81}\right)\left(\frac{243}{5}\right)$$

$$= 12 - \frac{24}{5}$$

$$= \frac{36}{5}.$$

We see, of course, that in the situation where D can be described in two different ways, the order in which we choose to set up and evaluate the double integral doesn't matter, and the same value results in either case.

The meaning of a double integral over a non-rectangular region, D, parallels the meaning over a rectangular region. In particular,

- $\iint_D f(x, y) \, dA$ tells us the volume of the solids the graph of f bounds above the xy-plane over the closed, bounded region D minus the volume of the solids the graph of f bounds below the xy-plane under the region D;

- $\frac{1}{A(D)} \iint_R f(x, y) \, dA$, where $A(D)$ is the area of D tells us the average value of the function f on D. If $f(x, y) \geq 0$ on D, we can interpret this average value of f on D as the height of the solid with base D and constant cross-sectional area D that has the same volume as the volume of the surface defined by f over D.

Activity 11.6.

Consider the double integral $\iint_D (4 - x - 2y) \, dA$, where D is the triangular region with vertices (0,0), (4,0), and (0,2).

(a) Write the given integral as an iterated integral of the form $\iint_D (4 - x - 2y) \, dy \, dx$. Draw a labeled picture of D with relevant cross sections.

(b) Write the given integral as an iterated integral of the form $\iint_D (4 - x - 2y) \, dx \, dy$. Draw a labeled picture of D with relevant cross sections.

(c) Evaluate the two iterated integrals from (a) and (b), and verify that they produce the same value. Give at least one interpretation of the meaning of your result.

◁

Activity 11.7.

Consider the iterated integral $\int_{x=3}^{x=5} \int_{y=-x}^{y=x^2} (4x + 10y) \, dy \, dx$.

(a) Sketch the region of integration, D, for which

$$\iint_D (4x + 10y) \, dA = \int_{x=3}^{x=5} \int_{y=-x}^{y=x^2} (4x + 10y) \, dy \, dx.$$

(b) Determine the equivalent iterated integral that results from integrating in the opposite order ($dx \, dy$, instead of $dy \, dx$). That is, determine the limits of integration for which

$$\iint_D (4x + 10y) \, dA = \int_{y=?}^{y=?} \int_{x=?}^{x=?} (4x + 10y) \, dx \, dy.$$

(c) Evaluate one of the two iterated integrals above. Explain what the value you obtained tells you.

(d) Set up and evaluate a single definite integral to determine the exact area of D, $A(D)$.

(e) Determine the exact average value of $f(x, y) = 4x + 10y$ over D.

◁

Activity 11.8.

Consider the iterated integral $\int_{x=0}^{x=4} \int_{y=x/2}^{y=2} e^{y^2} \, dy \, dx$.

(a) Explain why we cannot antidifferentiate e^{y^2} with respect to y, and thus are unable to evaluate the iterated integral $\int_{x=0}^{x=4} \int_{y=x/2}^{y=2} e^{y^2} \, dy \, dx$ using the Fundamental Theorem of Calculus.

(b) Sketch the region of integration, D, so that $\iint_D e^{y^2} \, dA = \int_{x=0}^{x=4} \int_{y=x/2}^{y=2} e^{y^2} \, dy \, dx$.

(c) Rewrite the given iterated integral in the opposite order, using $dA = dx \, dy$.

(d) Use the Fundamental Theorem of Calculus to evaluate the iterated integral you developed in (d). Write one sentence to explain the meaning of the value you found.

(e) What is the important lesson this activity offers regarding the order in which we set up an iterated integral?

Summary

- For a double integral $\iint_D f(x,y)\,dA$ over a non-rectangular region D, we enclose D in a rectangle R and then extend integrand f to a function F so that $F(x,y) = 0$ at all points in R outside of D and $F(x,y) = f(x,y)$ for all points in D. We then define $\iint_D f(x,y)\,dA$ to be equal to $\iint_R F(x,y)\,dA$.

- In an iterated double integral, the limits on the outer integral must be constants while the limits on the inner integral must be constants or in terms of only the remaining variable. In other words, an iterated double integral has one of the following forms (which result in the same value):

$$\int_{x=a}^{x=b} \int_{y=g_1(x)}^{y=g_2(x)} f(x,y)\,dy\,dx,$$

where $g_1 = g_1(x)$ and $g_2 = g_2(x)$ are functions of x only and the region D is described by the inequalities $g_1(x) \leq y \leq g_2(x)$ and $a \leq x \leq b$ or

$$\int_{y=c}^{y=d} \int_{x=h_1(y)}^{x=h_2(y)} f(x,y)\,dx\,dy,$$

where $h_1 = h_1(y)$ and $h_2 = h_2(y)$ are functions of y only and the region D is described by the inequalities $h_1(y) \leq x \leq h_2(y)$ and $c \leq y \leq d$.

Exercises

1. For each of the following iterated integrals, (a) sketch the region of integration, (b) write an equivalent iterated integral expression in the opposite order of integration, and (c) choose one of the two orders and evaluate the integral.

   (a) $\displaystyle\int_{x=0}^{x=1} \int_{y=x^2}^{y=x} xy\,dy\,dx$

   (b) $\displaystyle\int_{y=0}^{y=2} \int_{x=-\sqrt{4-y^2}}^{x=0} xy\,dx\,dy$

   (c) $\displaystyle\int_{x=0}^{x=1} \int_{y=x^4}^{y=x^{1/4}} x+y\,dy\,dx$

   (d) $\displaystyle\int_{y=0}^{y=2} \int_{x=y/2}^{x=2y} x+y\,dx\,dy$

2. The temperature at any point on a metal plate in the xy plane is given by $T(x, y) = 100 - 4x^2 - y^2$, where x and y are measured in inches and T in degrees Celsius. Consider the portion of the plate that lies on the region D that is the finite region that lies between the parabolas $x = y^2$ and $x = 3 - 2y^2$.

   (a) Construct a labeled sketch of the region D.

   (b) Set up an integrated integral whose value is $\iint_D T(x, y)\, dA$.

   (c) Set up an integrated integral whose value is $\iint_D T(x, y)\, dA$.

   (d) Use the Fundamental Theorem of Calculus to evaluate the integrals you determined in (b) and (c).

   (e) Determine the exact average temperature, $T_{\text{AVG}(D)}$, over the region D.

3. Consider the solid that is given by the following description: the base is the given region D, while the top is given by the surface $z = p(x, y)$. In each setting below, set up, but do not evaluate, an iterated integral whose value is the exact volume of the solid. Include a labeled sketch of D in each case.

   (a) D is the interior of the quarter circle of radius 2, centered at the origin, that lies in the second quadrant of the plane; $p(x, y) = 16 - x^2 - y^2$.

   (b) D is the finite region between the line $y = x + 1$ and the parabola $y = x^2$; $p(x, y) = 10 - x - 2y$.

   (c) D is the triangular region with vertices $(1, 1)$, $(2, 2)$, and $(2, 3)$; $p(x, y) = e^{-xy}$.

   (d) D is the region bounded by the y-axis, $y = 4$ and $x = \sqrt{y}$; $p(x, y) = \sqrt{1 + x^2 + y^2}$.

4. Consider the iterated integral $I = \displaystyle\int_{x=0}^{x=4} \int_{y=\sqrt{x}}^{y=2} \cos(y^3)\, dy\, dx$.

   (a) Sketch the region of integration.

   (b) Write an equivalent iterated integral with the order of integration reversed.

   (c) Choose one of the two orders of integration and evaluate the iterated integral you chose by hand. Explain the reasoning behind your choice.

   (d) Determine the exact average value of $\cos(y^3)$ over the region D that is determined by the iterated integral I.

11.4 Applications of Double Integrals

Motivating Questions

In this section, we strive to understand the ideas generated by the following important questions:

- If we have a mass density function for a lamina (thin plate), how does a double integral determine the mass of the lamina?

- How may a double integral be used to find the area between two curves?

- Given a mass density function on a lamina, how can we find the lamina's center of mass?

- What is a joint probability density function? How do we determine the probability of an event if we know a probability density function?

Introduction

So far, we have interpreted the double integral of a function f over a domain D in two different ways. First, $\iint_D f(x,y)\, dA$ tells us a difference of volumes – the volume the surface defined by f bounds above the xy-plane on D minus the volume the surface bounds below the xy-plane on D. In addition, $\frac{1}{A(D)} \iint_D f(x,y)\, dA$ determines the average value of f on D. In this section, we investigate several other applications of double integrals, using the integration process as seen in Preview Activity 11.4: we partition into small regions, approximate the desired quantity on each small region, then use the integral to sum these values exactly in the limit.

The following preview activity explores how a double integral can be used to determine the density of a thin plate with a mass density distribution. Recall that in single-variable calculus, we considered a similar problem and computed the mass of a one-dimensional rod with a mass-density distribution. There, as here, the key idea is that if density is constant, mass is the product of density and volume.

Preview Activity 11.4. Suppose that we have a flat, thin object (called a *lamina*) whose density varies across the object. We can think of the density on a lamina as a measure of mass per unit area. As an example, consider a circular plate D of radius 1 cm centered at the origin whose density δ varies depending on the distance from its center so that the density in grams per square centimeter at point (x, y) is

$$\delta(x, y) = 10 - 2(x^2 + y^2).$$

(a) Suppose that we partition the plate into subrectangles R_{ij}, where $1 \leq i \leq m$ and $1 \leq j \leq n$, of equal area ΔA, and select a point (x_{ij}^*, y_{ij}^*) in R_{ij} for each i and j.

What is the meaning of the quantity $\delta(x_{ij}^*, y_{ij}^*)\Delta A$?

(b) State a double Riemann sum that provides an approximation of the mass of the plate.

11.4. APPLICATIONS OF DOUBLE INTEGRALS

(c) Explain why the double integral

$$\iint_D \delta(x,y)\, dA$$

tells us the exact mass of the plate.

(d) Determine an iterated integral which, if evaluated, would give the exact mass of the plate. Do not actually evaluate the integral.[2]

Mass

Density is a measure of some quantity per unit area or volume. For example, we can measure the human population density of some region as the number of humans in that region divided by the area of that region. In physics, the mass density of an object is the mass of the object per unit area or volume. As suggested by Preview Activity 11.4, the following holds in general.

> If $\delta(x,y)$ describes the density of a lamina defined by a planar region D, then the **mass** of D is given by the double integral $\iint_D \delta(x,y)\, dA$.

Activity 11.9.

Let D be a half-disk lamina of radius 3 in quadrants IV and I, centered at the origin as shown in Figure 11.18. Assume the density at point (x, y) is given by $\delta(x, y) = x$. Find the exact mass of the lamina.

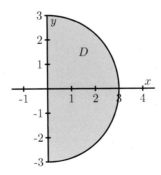

Figure 11.18: A half disk lamina.

---

[2]This integral is considerably easier to evaluate in polar coordinates, which we will learn more about in Section 11.5.

Area

If we consider the situation where the mass-density distribution is constant, we can also see how a double integral may be used to determine the area of a region. Assuming that $\delta(x,y) = 1$ over a closed bounded region D, where the units of δ are "mass per unit of area," it follows that $\iint_D 1\, dA$ is the mass of the lamina. But since the density is constant, the numerical value of the integral is simply the area.

As the following activity demonstrates, we can also see this fact by considering a three-dimensional solid whose height is always 1.

Activity 11.10.

Suppose we want to find the area of the bounded region D between the curves

$$y = 1 - x^2 \quad \text{and} \quad y = x - 1.$$

A picture of this region is shown in Figure 11.19.

(a) We know that the volume of a solid with constant height is given by the area of the base times the height. Hence, we may interpret the area of the region D as the volume of a solid with base D and of uniform height 1. Determine a double integral whose value is the area of D.

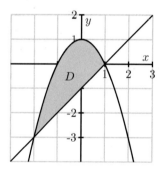

Figure 11.19: The graphs of $y = 1 - x^2$ and $y = x - 1$.

(b) Write an iterated integral whose value equals the double integral you found in (a).

(c) Use the Fundamental Theorem of Calculus to evaluate *only* the inner integral in the iterated integral in (b).

(d) After completing part (c), you should see a standard single area integral from calc II. Evaluate this remaining integral to find the exact area of D.

11.4. APPLICATIONS OF DOUBLE INTEGRALS

We now formally state the conclusion from our earlier discussion and Activity 11.10.

> Given a closed, bounded region D in the plane, the area of D, denoted $A(D)$, is given by the double integral
> $$A(D) = \iint_D 1 \, dA.$$

Center of Mass

The center of mass of an object is a point at which the object will balance perfectly. For example, the center of mass of a circular disk of uniform density is located at its center. For any object, if we throw it through the air, it will spin around its center of mass and behave as if all the mass is located at the center of mass.

In order to understand the role that integrals play in determining the center of a mass of an object with a nonuniform mass distribution, we start by finding the center of mass of a collection of N distinct point-masses in the plane.

Let $m_1, m_2, \ldots, m_N$ be N masses located in the plane. Think of these masses as connected by rigid rods of negligible weight from some central point (x, y). A picture with four masses is shown in Figure 11.20. Now imagine balancing this system by placing it on a thin pole at the point (x, y) perpendicular to the plane containing the masses. Unless the masses are perfectly balanced, the system will fall off the pole. The point $(\overline{x}, \overline{y})$ at which the system will balance perfectly is called the *center of mass* of the system. Our goal is to determine the center of mass of a system of discrete masses, then extend this to a continuous lamina.

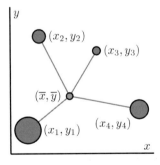

Figure 11.20: A center of mass $(\overline{x}, \overline{y})$ of four masses.

Each mass exerts a force (called a *moment*) around the lines $x = \overline{x}$ and $y = \overline{y}$ that causes the system to tilt in the direction of the mass. These moments are dependent on the mass and the distance from the given line. Let (x_1, y_1) be the location of mass m_1, (x_2, y_2) the location of mass m_2, etc. In order to balance perfectly, the moments in the x direction and in the y direction must be in equilibrium. We determine these moments and solve the resulting system to find the equilibrium point $(\overline{x}, \overline{y})$ at the center of mass.

The force that mass m_1 exerts to tilt the system from the line $y = \overline{y}$ is

$$m_1 g(\overline{y} - y_1),$$

where g is the gravitational constant. Similarly, the force mass m_2 exerts to tilt the system from the line $y = \overline{y}$ is

$$m_2 g(\overline{y} - y_2).$$

In general, the force that mass m_k exerts to tilt the system from the line $y = \overline{y}$ is

$$m_k g(\overline{y} - y_k).$$

For the system to balance, we need the forces to sum to 0, so that

$$\sum_{k=1}^{N} m_k g(\overline{y} - y_k) = 0.$$

Solving for $\overline{y}$, we find that

$$\overline{y} = \frac{\sum_{k=1}^{N} m_k y_k}{\sum_{k=1}^{N} m_k}.$$

A similar argument shows that

$$\overline{x} = \frac{\sum_{k=1}^{N} m_k x_k}{\sum_{k=1}^{N} m_k}.$$

The value $M_x = \sum_{k=1}^{N} m_k y_k$ is called the *total moment* with respect to the x-axis; $M_y = \sum_{k=1}^{N} m_k x_k$ is the *total moment* with respect to the y-axis. Hence, the respective quotients of the moments to the total mass, M, determines the center of mass of a point-mass system:

$$(\overline{x}, \overline{y}) = \left(\frac{M_y}{M}, \frac{M_x}{M} \right).$$

Now, suppose that rather than a point-mass system, we have a continuous lamina with a variable mass-density $\delta(x, y)$. We may estimate its center of mass by partitioning the lamina into mn subrectangles of equal area ΔA, and treating the resulting partitioned lamina as a point-mass system. In particular, we select a point (x_{ij}^*, y_{ij}^*) in the ijth subrectangle, and observe that the quanity

$$\delta(x_{ij}^*, y_{ij}^*) \Delta A$$

is density times area, so $\delta(x_{ij}^*, y_{ij}^*) \Delta A$ approximates the mass of the small portion of the lamina determined by the subrectangle R_{ij}.

We now treat $\delta(x_{ij}^*, y_{ij}^*) \Delta A$ as a point mass at the point (x_{ij}^*, y_{ij}^*). The coordinates $(\overline{x}, \overline{y})$ of the center of mass of these mn point masses are thus given by

$$\overline{x} = \frac{\sum_{j=1}^{n} \sum_{i=1}^{m} x_{ij}^* \delta(x_{ij}^*, y_{ij}^*) \Delta A}{\sum_{j=1}^{n} \sum_{i=1}^{m} \delta(x_{ij}^*, y_{ij}^*) \Delta A} \quad \text{and} \quad \overline{y} = \frac{\sum_{j=1}^{n} \sum_{i=1}^{m} y_{ij}^* \delta(x_{ij}^*, y_{ij}^*) \Delta A}{\sum_{j=1}^{n} \sum_{i=1}^{m} \delta(x_{ij}^*, y_{ij}^*) \Delta A}.$$

11.4. APPLICATIONS OF DOUBLE INTEGRALS

If we take the limit as m and n go to infinity, we obtain the exact center of mass $(\overline{x}, \overline{y})$ of the continuous lamina.

> The coordinates $(\overline{x}, \overline{y})$ of the **center of mass of a lamina** D with density $\delta = \delta(x, y)$ are given by
> $$\overline{x} = \frac{\iint_D x\delta(x,y)\, dA}{\iint_D \delta(x,y)\, dA} \quad \text{and} \quad \overline{y} = \frac{\iint_D y\delta(x,y)\, dA}{\iint_D \delta(x,y)\, dA}.$$

The numerators of $\overline{x}$ and $\overline{y}$ are called the respective *moments* of the lamina about the coordinate axes. Thus, the moment of a lamina D with density $\delta = \delta(x, y)$ about the y-axis is
$$M_y = \iint_D x\delta(x,y)\, dA$$
and the moment of D about the x-axis is
$$M_x = \iint_D y\delta(x,y)\, dA.$$

If M is the mass of the lamina, it follows that the center of mass is $(\overline{x}, \overline{y}) = \left(\frac{M_y}{M}, \frac{M_x}{M}\right)$.

Activity 11.11.

In this activity we determine integrals that represent the center of mass of a lamina D described by the triangular region bounded by the x-axis and the lines $x = 1$ and $y = 2x$ in the first quadrant if the density at point (x, y) is $\delta(x, y) = 6x + 6y + 6$. A picture of the lamina is shown in Figure 11.21.

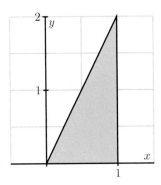

Figure 11.21: The lamina bounded by the x-axis and the lines $x = 1$ and $y = 2x$ in the first quadrant.

(a) Set up an iterated integral that represents the mass of the lamina.

(b) Assume the mass of the lamina is 14. Set up two iterated integrals that represent the coordinates of the center of mass of the lamina.

◁

Probability

Calculating probabilities is a very important application of integration in the physical, social, and life sciences. To understand the basics, consider the game of darts in which a player throws a dart at a board and tries to hit a particular target. Let us suppose that a dart board is in the form of a disk D with radius 10 inches. If we assume that a player throws a dart at random, and is not aiming at any particular point, then it is equally probable that the dart will strike any single point on the board. For instance, the probability that the dart will strike a particular 1 square inch region is $\frac{1}{100\pi}$, or the ratio of the area of the desired target to the total area of D (assuming that the dart thrower always hits the board itself at some point). Similarly, the probability that the dart strikes a point in the disk D_3 of radius 3 inches is given by the area of D_3 divided by the area of D. In other words, the probability that the dart strikes the disk D_3 is

$$\frac{9\pi}{100\pi} = \iint_{D_3} \frac{1}{100\pi} \, dA.$$

The integrand, $\frac{1}{100\pi}$, may be thought of as a *distribution function*, describing how the dart strikes are distributed across the board. In this case the distribution function is constant since we are assuming a uniform distribution, but we can easily envision situations where the distribution function varies. For example, if the player is fairly good and is aiming for the bulls eye (the center of D), then the distribution function f could be skewed toward the center, say

$$f(x, y) = K e^{-(x^2+y^2)}$$

for some constant positive K. If we assume that the player is consistent enough so that the dart always strikes the board, then the probability that the dart strikes the board somewhere is 1, and the distribution function f will have to satisfy[3]

$$\iint_D f(x, y) \, dA = 1.$$

For such a function f, the probability that the dart strikes in the disk D_1 of radius 1 would be

$$\iint_{D_1} f(x, y) \, dA.$$

Indeed, the probability that the dart strikes in any region R that lies within D is given by

$$\iint_R f(x, y) \, dA.$$

The preceding discussion highlights the general idea behind calculating probabilities. We assume we have a *joint probability density function* f, a function of two independent variables x and y defined on a domain D that satisfies the conditions

---

[3]This makes $K = \frac{1}{\pi(1-e^{-100})}$, which you can check.

11.4. APPLICATIONS OF DOUBLE INTEGRALS

- $f(x, y) \geq 0$ for all x and y in D,

- the probability that x is between some values a and b while y is between some values c and d is given by

$$\int_a^b \int_c^d f(x, y)\, dy\, dx,$$

- The probability that the point (x, y) is in D is 1, that is

$$\iint_D f(x, y)\, dA = 1. \qquad (11.2)$$

Note that it is possible that D could be an infinite region and the limits on the integral in Equation (11.2) could be infinite. When we have such a probability density function $f = f(x, y)$, the probability that the point (x, y) is in some region R contained in the domain D (the notation we use here is "$P((x, y) \in R)$") is determined by

$$P((x, y) \in R) = \iint_R f(x, y)\, dA.$$

Activity 11.12.

A firm manufactures smoke detectors. Two components for the detectors come from different suppliers – one in Michigan and one in Ohio. The company studies these components for their reliability and their data suggests that if x is the life span (in years) of a randomly chosen component from the Michigan supplier and y the life span (in years) of a randomly chosen component from the Ohio supplier, then the joint probability density function f might be given by

$$f(x, y) = e^{-x} e^{-y}.$$

(a) Theoretically, the components might last forever, so the domain D of the function f is the set D of all (x, y) such that $x \geq 0$ and $y \geq 0$. To show that f is a probability density function on D we need to demonstrate that

$$\iint_D f(x, y)\, dA = 1,$$

or that

$$\int_0^\infty \int_0^\infty f(x, y)\, dy\, dx = 1.$$

Use your knowledge of improper integrals to verify that f is indeed a probability density function.

(b) Assume that the smoke detector fails only if both of the supplied components fail. To determine the probability that a randomly selected detector will fail within one year, we will need to determine the probability that the life span of each component is between 0 and 1 years. Set up an appropriate iterated integral, and evaluate the integral to determine the probability.

(c) What is the probability that a randomly chosen smoke detector will fail between years 3 and 7?

(d) Suppose that the manufacturer determines that one of the components is more likely to fail than the other, and hence conjectures that the probability density function is instead $f(x,y) = Ke^{-x}e^{-2y}$. What is the value of K?

◁

Summary

- The mass of a lamina D with a mass density function $\delta = \delta(x,y)$ is $\iint_D \delta(x,y)\,dA$.

- The area of a region D in the plane has the same numerical value as the volume of a solid of uniform height 1 and base D, so the area of D is given by $\iint_D 1\,dA$.

- The center of mass, $(\overline{x}, \overline{y})$, of a continuous lamina with a variable density $\delta(x,y)$ is given by

$$\overline{x} = \frac{\iint_D x\delta(x,y)\,dA}{\iint_D \delta(x,y)\,dA} \quad \text{and} \quad \overline{y} = \frac{\iint_D y\delta(x,y)\,dA}{\iint_D \delta(x,y)\,dA}.$$

- Given a joint probability density function f is a function of two independent variables x and y defined on a domain D, if R is some subregion of D, then the probability that (x,y) is in R is given by

$$\iint_R f(x,y)\,dA.$$

Exercises

1. A triangular plate is bounded by the graphs of the equations $y = 2x$, $y = 4x$, and $y = 4$. The plate's density at (x,y) is given by $\delta(x,y) = 4xy^2 + 1$, measured in grams per square centimeter (and x and y are measured in centimeters).

   (a) Set up an iterated integral whose value is the mass of the plate. Include a labeled sketch of the region of integration. Why did you choose the order of integration you did?

   (b) Determine the mass of the plate.

   (c) Determine the exact center of mass of the plate. Draw and label the point you find on your sketch from (a).

   (d) What is the average density of the plate? Include units on your answer.

2. Let D be a half-disk lamina of radius 3 in quadrants IV and I, centered at the origin as in Activity 11.9. Assume the density at point (x,y) is equal to x.

   (a) Before doing any calculations, what do you expect the y-coordinate of the center of mass to be? Why?

11.4. APPLICATIONS OF DOUBLE INTEGRALS

(b) Set up iterated integral expressions which, if evaluated, will determine the exact center of mass of the lamina.

(c) Use appropriate technology to evaluate the integrals to find the center of mass numerically.

3. Let x denote the time (in minutes) that a person spends waiting in a checkout line at a grocery store and y the time (in minutes) that it takes to check out. Suppose the joint probability density for x and y is
$$f(x,y) = \frac{1}{8}e^{-x/4-y/2}.$$

(a) What is the exact probability that a person spends between 0 to 5 minutes waiting in line, and then 0 to 5 minutes waiting to check out?

(b) Set up, but do not evaluate, an iterated integral whose value determines the exact probability that a person spends at most 10 minutes total both waiting in line and checking out at this grocery store.

(c) Set up, but do not evaluate, an iterated integral expression whose value determines the exact probability that a person spends at least 10 minutes total both waiting in line and checking out, but not more than 20 minutes.

11.5 Double Integrals in Polar Coordinates

Motivating Questions

In this section, we strive to understand the ideas generated by the following important questions:

- What are the polar coordinates of a point in two-space?

- How do we convert between polar coordinates and rectangular coordinates?

- What is the area element in polar coordinates?

- How do we convert a double integral in rectangular coordinates to a double integral in polar coordinates?

Introduction

While we have naturally defined double integrals in the rectangular coordinate system, starting with domains that are rectangular regions, there are many of these integrals that are difficult, if not impossible, to evaluate. For example, consider the domain D that is the unit circle and $f(x, y) = e^{-x^2-y^2}$. To integrate f over D, we would use the iterated integral

$$\iint_D f(x,y)\, dA = \int_{x=-1}^{x=1} \int_{y=-\sqrt{1-x^2}}^{y=\sqrt{1-x^2}} e^{-x^2-y^2}\, dy\, dx.$$

For this particular integral, regardless of the order of integration, we are unable to find an antiderivative of the integrand; in addition, even if we were able to find an antiderivative, the inner limits of integration involve relatively complicated functions.

It is useful, therefore, to be able to translate to other coordinate systems where the limits of integration and evaluation of the involved integrals is simpler. In this section we provide a quick discussion of one such system – polar coordinates – and then introduce and investigate their ramifications for double integrals. The rectangular coordinate system allows us to consider domains and graphs relative to a rectangular grid. The polar coordinate system is an alternate coordinate system that allows us to consider domains less suited to rectangular coordinates, such as circles.

Preview Activity 11.5. The coordinates of a point determine its location. In particular, the rectangular coordinates of a point P are given by an ordered pair (x, y), where x is the (signed) distance the point lies from the y-axis to P and y is the (signed) distance the point lies from the x-axis to P. In polar coordinates, we locate the point by considering the distance the point lies from the origin, $(0, 0)$, and the angle the line segment from the origin to P forms with the positive x-axis.

(a) Determine the rectangular coordinates of the following points:

   i. The point P that lies 1 unit from the origin on the positive x-axis.

11.5. DOUBLE INTEGRALS IN POLAR COORDINATES

ii. The point Q that lies 2 units from the origin and such that $\overline{OQ}$ makes an angle of $\frac{\pi}{2}$ with the positive x-axis, where O is the origin, $(0, 0)$.

iii. The point R that lies 3 units from the origin such that $\overline{OR}$ makes an angle of $\frac{2\pi}{3}$ with the positive x-axis.

(b) Part (a) indicates that the two pieces of information completely determine the location of a point: either the traditional (x, y) coordinates, or alternately, the distance r from the point to the origin along with the angle θ that the line through the origin and the point makes with the positive x-axis. We write "(r, θ)" to denote the point's location in its polar coordinate representation. Find polar coordinates for the points with the given rectangular coordinates.

i. $(0, -1)$ ii. $(-2, 0)$ iii. $(-1, 1)$

(c) For each of the following points whose coordinates are given in polar form, determine the rectangular coordinates of the point.

i. $(5, \frac{\pi}{4})$ ii. $(2, \frac{5\pi}{6})$ iii. $(\sqrt{3}, \frac{5\pi}{3})$

⋈

A Quick Overview of Polar Coordinates

The rectangular coordinate system is best suited for graphs and regions that are naturally considered over a rectangular grid. The polar coordinate system is an alternative that offers good options for functions and domains that have more circular characteristics. A point P in rectangular coordinates that is described by an ordered pair (x, y), where x is the displacement from P to the y-axis and y is the displacement from P to the x-axis, as seen in Preview Activity 11.5, can also be described with polar coordinates (r, θ), where r is the distance from P to the origin and θ is the angle formed by the line segment $\overline{OP}$ and the positive x-axis, as shown in Figure 11.22. Trigonometry and the Pythagorean Theorem allow for straightforward conversion from rectangular to polar, and vice versa.

> **Converting from rectangular to polar.** If we are given the rectangular coordinates (x, y) of a point P, then the polar coordinates (r, θ) of P satisfy
>
> $$r = \sqrt{x^2 + y^2} \quad \text{and} \quad \tan(\theta) = \frac{y}{x}, \text{ assuming } x \neq 0.$$
>
> **Converting from polar to rectangular.** If we are given the polar coordinates (r, θ) of a point P, then the rectangular coordinates (x, y) of P satisfy
>
> $$x = r\cos(\theta) \quad \text{and} \quad y = r\sin(\theta).$$

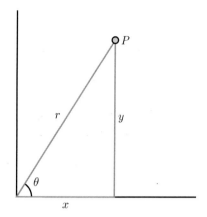

Figure 11.22: The polar coordinates of a point.

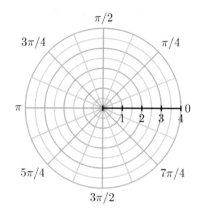

Figure 11.23: The polar coordinate grid.

We can draw graphs of curves in polar coordinates in a similar way to how we do in rectangular coordinates. However, when plotting in polar coordinates, we use a grid that considers changes in angles and changes in distance from the origin. In particular, the angles θ and distances r partition the plane into small wedges as shown in Figure 11.23.

Activity 11.13.

Most polar graphing devices[4] can plot curves in polar coordinates of the form $r = f(\theta)$. Use such a device to complete this activity.

(a) Before plotting the polar curve $r = 1$, think about what shape it should have, in light of how r is connected to x and y. Then use appropriate technology to draw the graph and test your intuition.

(b) The equation $\theta = 1$ does not define r as a function of θ, so we can't graph this equation on many polar plotters. What do you think the graph of the polar curve $\theta = 1$ looks like? Why?

(c) Before plotting the polar curve $r = \theta$, what do you think the graph looks like? Why? Use technology to plot the curve and compare your intuition.

(d) What about the curve $r = \sin(\theta)$? After plotting this curve, experiment with others of your choosing and think about why the curves look the way they do.

◁

---

[4]You can use your calculator in POL mode, or a web applet such as http://webspace.ship.edu/msrenault/ggb/polar_grapher.html

11.5. DOUBLE INTEGRALS IN POLAR COORDINATES

Integrating in Polar Coordinates

Consider the double integral

$$\iint_D e^{x^2+y^2} \, dA,$$

where D is the unit disk. While we cannot directly evaluate this integral in rectangular coordinates, a change to polar coordinates will convert it to one we can easily evaluate.

We have seen how to evaluate a double integral $\iint_D f(x,y) \, dA$ as an iterated integral of the form

$$\int_a^b \int_{g_1(x)}^{g_2(x)} f(x,y) \, dy \, dx$$

in rectangular coordinates, because we know that $dA = dy \, dx$ in rectangular coordinates. To make the change to polar coordinates, we not only need to represent the variables x and y in polar coordinates, but we also must understand how to write the area element, dA, in polar coordinates. That is, we must determine how the area element dA can be written in terms of dr and $d\theta$ in the context of polar coordinates. We address this question in the following activity.

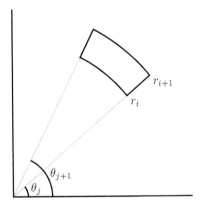

Figure 11.24: A polar rectangle.

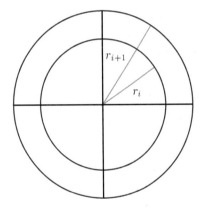

Figure 11.25: An annulus.

Activity 11.14.

Consider a polar rectangle R, with r between r_i and r_{i+1} and θ between θ_j and θ_{j+1} as shown in Figure 11.24. Let $\Delta r = r_{i+1} - r_i$ and $\Delta \theta = \theta_{j+1} - \theta_j$. Let ΔA be the area of this region.

(a) Explain why the area ΔA in polar coordinates is not $\Delta r \, \Delta \theta$.

(b) Now find ΔA by the following steps:

    i. Find the area of the annulus (the washer-like region) between r_i and r_{i+1}, as shown at right in Figure 11.25. This area will be in terms of r_i and r_{i+1}.

ii. Observe that the region R is only a portion of the annulus, so the area ΔA of R is only a fraction of the area of the annulus. For instance, if $\theta_{i+1} - \theta_i$ were $\frac{\pi}{4}$, then the resulting wedge would be

$$\frac{\frac{\pi}{4}}{2\pi} = \frac{1}{8}$$

of the entire annulus. In this more general context, using the wedge between the two noted angles, what fraction of the area of the annulus is the area ΔA?

iii. Write an expression for ΔA in terms of r_i, r_{i+1}, θ_j, and θ_{j+1}.

iv. Finally, write the area ΔA in terms of r_i, r_{i+1}, Δr, and $\Delta \theta$, where each quantity appears only once in the expression. (Hint: Think about how to factor a difference of squares.)

(c) As we take the limit as Δr and $\Delta \theta$ go to 0, Δr becomes dr, $\Delta \theta$ becomes $d\theta$, and ΔA becomes dA, the area element. Using your work in (iv), write dA in terms of r, dr, and $d\theta$.

◁

From the result of Activity 11.14, we see when we convert an integral from rectangular coordinates to polar coordinates, we must not only convert x and y to being in terms of r and θ, but we also have to change the area element to $dA = r\, dr\, d\theta$ in polar coordinates. In other words, given a double integral $\iint_D f(x, y)\, dA$ in rectangular coordinates, to write a corresponding iterated integral in polar coordinates, we replace x with $r\cos(\theta)$, y with $r\sin(\theta)$ and dA with $r\, dr\, d\theta$. Of course, we need to describe the region D in polar coordinates as well. To summarize:

> The double integral $\iint_D f(x, y)\, dA$ in rectangular coordinates can be converted to a double integral in polar coordinates as $\iint_D f(r\cos(\theta), r\sin(\theta))\, r\, dr\, d\theta$.

Example 11.2. Let $f(x, y) = e^{x^2+y^2}$ on the disk $D = \{(x, y) : x^2 + y^2 \leq 1\}$. We will evaluate $\iint_D f(x, y)\, dA$.

In rectangular coordinates the double integral $\iint_D f(x, y)\, dA$ can be written as the iterated integral

$$\iint_D f(x, y)\, dA = \int_{x=-1}^{x=1} \int_{y=-\sqrt{1-x^2}}^{y=\sqrt{1-x^2}} e^{x^2+y^2}\, dy\, dx.$$

We cannot evaluate this iterated integral, because $e^{x^2+y^2}$ does not have an elementary antiderivative with respect to either x or y. However, since $r^2 = x^2 + y^2$ and the region D is circular, it is natural to wonder whether converting to polar coordinates will allow us to evaluate the new integral. To do so, we replace x with $r\cos(\theta)$, y with $r\sin(\theta)$, and $dy\, dx$ with $r\, dr\, d\theta$ to obtain

$$\iint_D f(x, y)\, dA = \iint_D e^{r^2}\, r\, dr\, d\theta.$$

11.5. DOUBLE INTEGRALS IN POLAR COORDINATES

The disc D is described in polar coordinates by the constraints $0 \leq r \leq 1$ and $0 \leq \theta \leq 2\pi$. Therefore, it follows that

$$\iint_D e^{r^2} \, r \, dr \, d\theta = \int_{\theta=0}^{\theta=2\pi} \int_{r=0}^{r=1} e^{r^2} \, r \, dr \, d\theta.$$

We can evaluate the resulting iterated polar integral as follows:

$$\int_{\theta=0}^{\theta=2\pi} \int_{r=0}^{r=1} e^{r^2} \, r \, dr \, d\theta = \int_{\theta=0}^{2\pi} \left(\frac{1}{2} e^{r^2} \Big|_{r=0}^{r=1} \right) d\theta$$

$$= \frac{1}{2} \int_{\theta=0}^{\theta=2\pi} (e-1) \, d\theta$$

$$= \frac{1}{2}(e-1) \int_{\theta=0}^{\theta=2\pi} d\theta$$

$$= \frac{1}{2}(e-1) [\theta] \Big|_{\theta=0}^{\theta=2\pi}$$

$$= \pi(e-1).$$

While there is no firm rule for when polar coordinates can or should be used, they are a natural alternative anytime the domain of integration may be expressed simply in polar form, and/or when the integrand involves expressions such as $\sqrt{x^2 + y^2}$.

Activity 11.15.

Let $f(x, y) = x + y$ and $D = \{(x, y) : x^2 + y^2 \leq 4\}$.

(a) Write the double integral of f over D as an iterated integral in rectangular coordinates.

(b) Write the double integral of f over D as an iterated integral in polar coordinates.

(c) Evaluate one of the iterated integrals. Why is the final value you found not surprising?

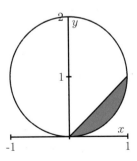

Figure 11.26: The graphs of $y = x$ and $x^2 + (y-1)^2 = 1$, for use in Activity 11.16.

Activity 11.16.

Consider the circle given by $x^2 + (y-1)^2 = 1$ as shown in Figure 11.26.

(a) Determine a polar curve in the form $r = f(\theta)$ that traces out the circle $x^2 + (y-1)^2 = 1$.

(b) Find the exact average value of $g(x, y) = \sqrt{x^2 + y^2}$ over the interior of the circle $x^2 + (y-1)^2 = 1$.

(c) Find the volume under the surface $h(x, y) = x$ over the region D, where D is the region bounded above by the line $y = x$ and below by the circle.

(d) Explain why in both (b) and (c) it is advantageous to use polar coordinates.

◁

Summary

- The polar representation of a point P is the ordered pair (r, θ) where r is the distance from the origin to P and θ is the angle the ray through the origin and P makes with the positive x-axis.

- The polar coordinates r and θ of a point (x, y) in rectangular coordinates satisfy
$$r = \sqrt{x^2 + y^2} \quad \text{and} \quad \tan(\theta) = \frac{y}{x};$$
the rectangular coordinates x and y of a point (r, θ) in polar coordinates satisfy
$$x = r\cos(\theta) \quad \text{and} \quad y = r\sin(\theta).$$

- The area element dA in polar coordinates is determined by the area of a slice of an annulus and is given by
$$dA = r\, dr\, d\theta.$$

- To convert the double integral $\iint_D f(x, y)\, dA$ to an iterated integral in polar coordinates, we substitute $r\cos(\theta)$ for x, $r\sin(\theta)$ for y, and $r\, dr\, d\theta$ for dA to obtain the iterated integral
$$\iint_D f(r\cos(\theta), r\sin(\theta))\, r\, dr\, d\theta.$$

Exercises

1. Consider the iterated integral $I = \displaystyle\int_{-3}^{0} \int_{-\sqrt{9-y^2}}^{0} \frac{y}{x^2 + y^2 + 1}\, dx\, dy$.

    (a) Sketch (and label) the region of integration.

    (b) Convert the given iterated integral to one in polar coordinates.

    (c) Evaluate the iterated integral in (b).

(d) State one possible interpretation of the value you found in (c).

2. Let D be the region that lies inside the unit circle in the plane.

   (a) Set up and evaluate an iterated integral in polar coordinates whose value is the area of D.

   (b) Determine the exact average value of $f(x,y) = y$ over the upper half of D.

   (c) Find the exact center of mass of the lamina over the portion of D that lies in the first quadrant and has its mass density distribution given by $\delta(x,y) = 1$. (Before making any calculations, where do you expect the center of mass to lie? Why?)

   (d) Find the exact volume of the solid that lies under the surface $z = 8 - x^2 - y^2$ and over the unit disk, D.

3. For each of the following iterated integrals, (a) sketch and label the region of integration, (b) convert the integral to the other coordinate system (if given in polar, to rectangular; if given in rectangular, to polar), and (c) choose one of the two iterated integrals to evaluate exactly.

   (a) $\displaystyle\int_{\pi}^{3\pi/2} \int_{0}^{3} r^3 \, dr \, d\theta$

   (b) $\displaystyle\int_{0}^{2} \int_{-\sqrt{1-(x-1)^2}}^{\sqrt{1-(x-1)^2}} \sqrt{x^2 + y^2} \, dy \, dx$

   (c) $\displaystyle\int_{0}^{\pi/2} \int_{0}^{\sin(\theta)} r\sqrt{1 - r^2} \, dr \, d\theta$.

   (d) $\displaystyle\int_{0}^{\sqrt{2}/2} \int_{y}^{\sqrt{1-y^2}} \cos(x^2 + y^2) \, dx \, dy$.

11.6 Surfaces Defined Parametrically and Surface Area

Motivating Questions

In this section, we strive to understand the ideas generated by the following important questions:

- What is a parameterization of a surface?
- How do we find the surface area of a parametrically defined surface?

Introduction

We have now studied at length how curves in space can be defined parametrically by functions of the form $\mathbf{r}(t) = \langle x(t), y(t), z(t) \rangle$, and surfaces can be represented by functions $z = f(x, y)$. In what follows, we will see how we can also define surfaces parametrically. A one-dimensional curve in space results from a vector function that relies upon one parameter, so a two-dimensional surface naturally involves the use of two parameters. If $x = x(s, t)$, $y = y(s, t)$, and $z = z(s, t)$ are functions of independent parameters s and t, then the terminal points of all vectors of the form

$$\mathbf{r}(s, t) = x(s, t)\mathbf{i} + y(s, t)\mathbf{j} + z(s, t)\mathbf{k}$$

form a surface in space. The equations $x = x(s, t)$, $y = y(s, t)$, and $z = z(s, t)$ are the *parametric equations* for the surface, or a *parametrization* of the surface. In Preview Activity 11.6 we investigate how to parameterize a cylinder and a cone.

Preview Activity 11.6. Recall the standard parameterization of the unit circle that is given by

$$x(t) = \cos(t) \quad \text{and} \quad y(t) = \sin(t),$$

where $0 \leq t \leq 2\pi$.

(a) Determine a parameterization of the circle of radius 1 in $\mathbb{R}^3$ that has its center at $(0, 0, 1)$ and lies in the plane $z = 1$.

(b) Determine a parameterization of the circle of radius 1 in 3-space that has its center at $(0, 0, -1)$ and lies in the plane $z = -1$.

(c) Determine a parameterization of the circle of radius 1 in 3-space that has its center at $(0, 0, 5)$ and lies in the plane $z = 5$.

(d) Taking into account your responses in (a), (b), and (c), describe the graph that results from the set of parametric equations

$$x(s, t) = \cos(t), \quad y(s, t) = \sin(t), \quad \text{and} \quad z(s, t) = s,$$

where $0 \leq t \leq 2\pi$ and $-5 \leq s \leq 5$. Explain your thinking.

(e) Just as a cylinder can be viewed as a "stack" of circles of constant radius, a cone can be viewed as a stack of circles with varying radius. Modify the parametrizations of the circles above in order to construct the parameterization of a cone whose vertex lies at the origin, whose base radius is 4, and whose height is 3, where the base of the cone lies in the plane $z = 3$. Use appropriate technology[5] to plot the parametric equations you develop. (Hint: The cross sections parallel to the xz plane are circles, with the radii varying linearly as z increases.)

⋈

Parametric Surfaces

In a single-variable setting, any function may have its graph expressed parametrically. For instance, given $y = g(x)$, by considering the parameterization $\langle t, g(t) \rangle$ (where t belongs to the domain of g), we generate the same curve. What is more important is that certain curves that are not functions may be represented parametrically; for instance, the circle (which cannot be represented by a single function) can be parameterized by $\langle \cos(t), \sin(t) \rangle$, where $0 \leq t \leq 2\pi$.

In the same way, in a two-variable setting, the surface $z = f(x, y)$ may be expressed parametrically by considering

$$\langle x(s,t), y(s,t), z(s,t) \rangle = \langle s, t, f(s,t) \rangle,$$

where (s, t) varies over the entire domain of f. Therefore, any familiar surface that we have studied so far can be generated as a parametric surface. But what is more powerful is that there are surfaces that cannot be generated by a single function $z = f(x, y)$ (such as the unit sphere), but that can be represented parametrically. We now consider an important example.

Example 11.3. Consider the torus (or doughnut) shown in Figure 11.27.

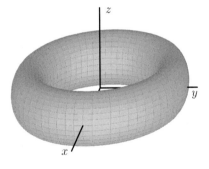

Figure 11.27: A torus

To find a parametrization of this torus, we recall our work in Preview Activity 11.6. There, we saw that a circle of radius r that has its center at the point $(0, 0, z_0)$ and is contained in the

---

[5]e.g., http://www.flashandmath.com/mathlets/multicalc/paramrec/surf_graph_rectan.html

horizontal plane $z = z_0$, as shown in Figure 11.28, can be parametrized using the vector-valued function **r** defined by

$$\mathbf{r}(t) = r\cos(t)\mathbf{i} + r\sin(t)\mathbf{j} + z_0\mathbf{k}$$

where $0 \leq t \leq 2\pi$.

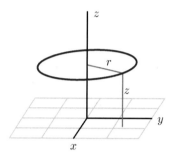

Figure 11.28: A circle in a horizontal plane centered at $(0, 0, z_0)$.

To obtain the torus in Figure 11.27, we begin with a circle of radius a in the xz-plane centered at $(b, 0)$, as shown on the left of Figure 11.29. We may parametrize the points on this circle, using the parameter s, by using the equations

$$x(s) = b + a\cos(s) \quad \text{and} \quad z(s) = a\sin(s),$$

where $0 \leq s \leq 2\pi$.

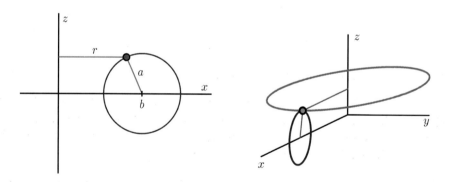

Figure 11.29: Revolving a circle to obtain a torus.

Let's focus our attention on one point on this circle, such as the indicated point, which has coordinates $(x(s), 0, z(s))$ for a fixed value of the parameter s. When this point is revolved about the z-axis, we obtain a circle contained in a horizontal plane centered at $(0, 0, z(s))$ and having radius $x(s)$, as shown on the right of Figure 11.29. If we let t be the new parameter that generates the circle for the rotation about the z-axis, this circle may be parametrized by

$$\mathbf{r}(s, t) = x(s)\cos(t)\mathbf{i} + x(s)\sin(t)\mathbf{j} + z(s)\mathbf{k}.$$

11.6. SURFACES DEFINED PARAMETRICALLY AND SURFACE AREA

Now using our earlier parametric equations for $x(s)$ and $z(s)$ for the original smaller circle, we have an overall parameterization of the torus given by

$$\mathbf{r}(s,t) = (b + a\cos(s))\cos(t)\mathbf{i} + (b + a\cos(s))\sin(t)\mathbf{j} + a\sin(s)\mathbf{k}.$$

To trace out the entire torus, we require that the parameters vary through the values $0 \le s \le 2\pi$ and $0 \le t \le 2\pi$.

Activity 11.17.

In this activity, we seek a parametrization of the sphere of radius R centered at the origin, as shown on the left in Figure 11.30. Notice that this sphere may be obtained by revolving a half-circle contained in the xz-plane about the z-axis, as shown on the right.

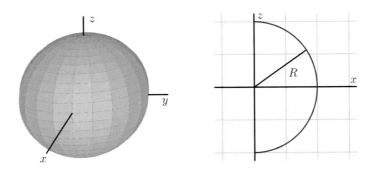

Figure 11.30: A sphere obtained by revolving a half-circle.

(a) Begin by writing a parametrization of this half-circle using the parameter s:

$$x(s) = \ldots, \qquad z(s) = \ldots.$$

Be sure to state the domain of the parameter s.

(b) By revolving the points on this half-circle about the z-axis, obtain a parametrization $\mathbf{r}(s,t)$ of the points on the sphere of radius R. Be sure to include the domain of both parameters s and t. (Hint: What is the radius of the circle obtained when revolving a point on the half-circle around the z axis?)

(c) Draw the surface defined by your parameterization with appropriate technology[6].

◁

The Surface Area of Parametrically Defined Surfaces

Recall that a differentiable function is locally linear – that is, if we zoom in on the surface around a point, the surface looks like its tangent plane. We now exploit this idea in order to determine

---

[6]e.g., http://web.monroecc.edu/manila/webfiles/calcNSF/JavaCode/CalcPlot3D.htm or http://www.flashandmath.com/mathlets/multicalc/paramrec/surf_graph_rectan.html

the surface area generated by a parametrization $\langle x(s,t), y(s,t), z(s,t) \rangle$. The basic idea is a familiar one: we will subdivide the surface into small pieces, in the approximate shape of small parallelograms, and thus estimate the entire the surface area by adding the areas of these approximation parallelograms. Ultimately, we use an integral to sum these approximations and determine the exact surface area.

Let
$$\mathbf{r}(s,t) = x(s,t)\mathbf{i} + y(s,t)\mathbf{j} + z(s,t)\mathbf{k}$$
define a surface over a rectangular domain $a \leq s \leq b$ and $c \leq t \leq d$. As a function of two variables, s and t, it is natural to consider the two partial derivatives of the vector-valued function $\mathbf{r}$, which we define by

$$\mathbf{r}_s(s,t) = x_s(s,t)\mathbf{i} + y_s(s,t)\mathbf{j} + z_s(s,t)\mathbf{k} \quad \text{and} \quad \mathbf{r}_t(s,t) = x_t(s,t)\mathbf{i} + y_t(s,t)\mathbf{j} + z_t(s,t)\mathbf{k}.$$

In the usual way, we slice the domain into small rectangles. In particular, we partition the interval $[a,b]$ into m subintervals of length $\Delta s = \frac{b-a}{n}$ and let $s_0, s_1, \ldots, s_m$ be the endpoints of these subintervals, where $a = s_0 < s_1 < s_2 < \cdots < s_m = b$. Also partition the interval $[c,d]$ into n subintervals of equal length $\Delta t = \frac{d-c}{n}$ and let $t_0, t_1, \ldots, t_n$ be the endpoints of these subintervals, where $c = t_0 < t_1 < t_2 < \cdots < t_n = d$. These two partitions create a partition of the rectangle $R = [a,b] \times [c,d]$ in st-coordinates into mn sub-rectangles R_{ij} with opposite vertices (s_{i-1}, t_{j-1}) and (s_i, t_j) for i between 1 and m and j between 1 and n. These rectangles all have equal area $\Delta A = \Delta s \cdot \Delta t$.

Now we want to think about the small piece of area on the surface itself that lies above one of these small rectangles in the domain. Observe that if we increase s by a small amount Δs from the point (s_{i-1}, t_{j-1}) in the domain, then $\mathbf{r}$ changes by approximately $\mathbf{r}_s(s_{i-1}, t_{j-1})\Delta s$. Similarly, if we increase t by a small amount Δt from the point (s_{i-1}, t_{j-1}), then $\mathbf{r}$ changes by approximately $\mathbf{r}_t(s_{i-1}, t_{j-1})\Delta t$. So we can approximate the surface defined by $\mathbf{r}$ on the st-rectangle $[s_{i-1}, s_i] \times [t_{j-1}, t_j]$ with the parallelogram determined by the vectors $\mathbf{r}_s(s_{i-1}, t_{j-1})\Delta s$ and $\mathbf{r}_t(s_{i-1}, t_{j-1})\Delta t$, as seen in Figure 11.31.

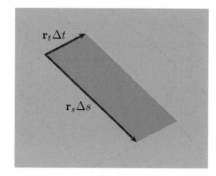

Figure 11.31: Approximation surface area with a parallelogram.

Say that the small parallelogram has area S_{ij}. If we can find its area, then all that remains is to sum the areas of all of the generated parallelograms and take a limit. Recall from our earlier work

11.6. SURFACES DEFINED PARAMETRICALLY AND SURFACE AREA

in the course that given two vectors **u** and **v**, the area of the parallelogram spanned by **u** and **v** is given by the magnitude of their cross product, $|\mathbf{u} \times \mathbf{v}|$. In the present context, it follows that the area, S_{ij}, of the parallelogram determined by the vectors $\mathbf{r}_s(s_{i-1}, t_{j-1})\Delta s$ and $\mathbf{r}_t(s_{i-1}, t_{j-1})\Delta t$ is

$$S_{ij} = |(\mathbf{r}_s(s_{i-1}, t_{j-1})\Delta s) \times (\mathbf{r}_t(s_{i-1}, t_{j-1})\Delta t)| = |\mathbf{r}_s(s_{i-1}, t_{j-1}) \times \mathbf{r}_t(s_{i-1}, t_{j-1})|\Delta s \Delta t, \quad (11.3)$$

where the latter equality holds from standard properties of the cross product and length.

We sum the surface area approximations from Equation (11.3) over all sub-rectangles to obtain an estimate for the total surface area, S, given by

$$S \approx \sum_{i=1}^{m} \sum_{j=1}^{n} |\mathbf{r}_s(s_{i-1}, t_{j-1}) \times \mathbf{r}_t(s_{i-1}, t_{j-1})|\Delta s \Delta t.$$

Taking the limit as $m, n \to \infty$ shows that the surface area of the surface defined by **r** over the domain D is given as follows.

Let $\mathbf{r}(s, t) = \langle x(s, t), y(s, t), z(s, t) \rangle$ be a parameterization of a smooth surface over a domain D. The **area of the surface** defined by **r** on D is given by

$$S = \iint_D |\mathbf{r}_s \times \mathbf{r}_t| \, dA. \quad (11.4)$$

Activity 11.18.

Consider the cylinder with radius a and height h defined parametrically by

$$\mathbf{r}(s, t) = a\cos(s)\mathbf{i} + a\sin(s)\mathbf{j} + t\mathbf{k}$$

for $0 \leq s \leq 2\pi$ and $0 \leq t \leq h$, as shown in Figure 11.32.

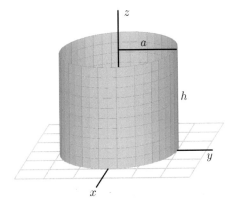

Figure 11.32: A cylinder.

(a) Set up an iterated integral to determine the surface area of this cylinder.

(b) Evaluate the iterated integral.

(c) Recall that one way to think about the surface area of a cylinder is to cut the cylinder horizontally and find the perimeter of the resulting cross sectional circle, then multiply by the height. Calculate the surface area of the given cylinder using this alternate approach, and compare your work in (b).

◁

As we noted earlier, we can take any surface $z = f(x, y)$ and generate a corresponding parameterization for the surface by writing $\langle s, t, f(s, t) \rangle$. Hence, we can use our recent work with parametrically defined surfaces to find the surface area that is generated by a function $f = f(x, y)$ over a given domain.

Activity 11.19.

Let $z = f(x, y)$ define a smooth surface, and consider the corresponding parameterization $\mathbf{r}(s, t) = \langle s, t, f(s, t) \rangle$.

(a) Let D be a region in the domain of f. Using Equation 11.4, show that the area, S, of the surface defined by the graph of f over D is

$$S = \iint_D \sqrt{(f_x(x, y))^2 + (f_y(x, y))^2 + 1} \, dA.$$

(b) Use the formula developed in (a) to calculate the area of the surface defined by $f(x, y) = \sqrt{4 - x^2}$ over the rectangle $D = [-2, 2] \times [0, 3]$.

(c) Observe that the surface of the solid describe in (b) is half of a circular cylinder. Use the standard formula for the surface area of a cylinder to calculate the surface area in a different way, and compare your result from (b).

◁

Summary

- A parameterization of a curve describes the coordinates of a point on the curve in terms of a single parameter t, while a parameterization of a surface describes the coordinates of points on the surface in terms of two independent parameters.

- If $\mathbf{r}(s, t) = \langle x(s, t), y(s, t), z(s, t) \rangle$ describes a smooth surface in 3-space on a domain D, then the area, S, of that surface is given by

$$S = \iint_D |\mathbf{r}_s \times \mathbf{r}_t| \, dA.$$

11.6. SURFACES DEFINED PARAMETRICALLY AND SURFACE AREA

Exercises

1. Consider the ellipsoid given by the equation
$$\frac{x^2}{16} + \frac{y^2}{25} + \frac{z^2}{9} = 1.$$

   In Activity 11.17, we found that a parameterization of the sphere S of radius R centered at the origin is
$$x(r,s) = R\cos(s)\cos(t), \quad y(s,t) = R\cos(s)\sin(t), \text{ and } z(s,t) = R\sin(s)$$
   for $-\frac{\pi}{2} \leq s \leq \frac{\pi}{2}$ and $0 \leq t \leq 2\pi$.

   (a) Let (x, y, z) be a point on the ellipsoid and let $X = \frac{x}{4}$, $Y = \frac{y}{5}$, and $Z = \frac{z}{3}$. Show that (X, Y, Z) lies on the sphere S. Hence, find a parameterization of S in terms of X, Y, and Z as functions of s and t.

   (b) Use the result of part (a) to find a parameterization of the ellipse in terms of x, y, and z as functions of s and t. Check your parametrization by substituting x, y, and z into the equation of the ellipsoid. Then check your work by plotting the surface defined by your parameterization with appropriate technology[7].

2. In this exercise, we explore how to use a parametrization and iterated integral to determine the surface area of a sphere.

   (a) Set up an iterated integral whose value is the portion of the surface area of a sphere of radius R that lies in the first octant (see the parameterization you developed in Activity 11.17).

   (b) Then, evaluate the integral to calculate the surface area of this portion of the sphere.

   (c) By what constant must you multiply the value determined in (b) in order to find the total surface area of the entire sphere.

   (d) Finally, compare your result to the standard formula for the surface area of sphere.

3. Consider the plane generated by $z = f(x, y) = 24 - 2x - 3y$ over the region $D = [0, 2] \times [0, 3]$.

   (a) Sketch a picture of the overall solid generated by the plane over the given domain.

   (b) Determine a parameterization $\mathbf{r}(s, t)$ for the plane over the domain D.

   (c) Use Equation 11.4 to determine the surface area generated by f over the domain D.

   (d) Observe that the vector $\mathbf{u} = \langle 2, 0, -4 \rangle$ points from $(0, 0, 24)$ to $(2, 0, 20)$ along one side of the surface generated by the plane f over D. Find the vector $\mathbf{v}$ such that $\mathbf{u}$ and $\mathbf{v}$ together span the parallelogram that represents the surface defined by f over D, and hence compute $|\mathbf{u} \times \mathbf{v}|$. What do you observe about the value you find?

---

[7]e.g., http://web.monroecc.edu/manila/webfiles/calcNSF/JavaCode/CalcPlot3D.htm or http://www.flashandmath.com/mathlets/multicalc/paramrec/surf_graph_rectan.html

4. A cone with base radius a and height h can be realized as the surface defined by $z = \frac{h}{a}\sqrt{x^2 + y^2}$, where a and h are positive.

   (a) Find a parameterization of the cone described by $z = \frac{h}{a}\sqrt{x^2 + y^2}$. (Hint: Compare to the parameterization of a cylinder as seen in Activity 11.18.)

   (b) Set up an iterated integral to determine the surface area of this cone.

   (c) Evaluate the iterated integral to find a formula for the lateral surface area of a cone of height h and base a.

11.7 Triple Integrals

> **Motivating Questions**
>
> *In this section, we strive to understand the ideas generated by the following important questions:*
>
> - How are a triple Riemann sum and the corresponding triple integral of a continuous function $f = f(x, y, z)$ defined?
>
> - What are two things the triple integral of a function can tell us?

Introduction

We have now learned that we define the double integral of a continuous function $f = f(x, y)$ over a rectangle $R = [a, b] \times [c, d]$ as a limit of a double Riemann sum, and that these ideas parallel the single-variable integral of a function $g = g(x)$ on an interval $[a, b]$. Moreover, this double integral has natural interpretations and applications, and can even be considered over non-rectangular regions, D. For instance, given a continuous function f over a region D, the average value of f, $f_{\text{AVG}(D)}$, is given by

$$f_{\text{AVG}(R)} = \frac{1}{A(D)} \iint_D f(x, y)\, dA,$$

where $A(D)$ is the area of D. Likewise, if $\delta(x, y)$ describes a mass density function on a lamina over D, the mass, M, of the lamina is given by

$$M = \iint_D \delta(x, y)\, dA.$$

It is natural to wonder if it is possible to extend these ideas of double Riemann sums and double integrals for functions of two variables to triple Riemann sums and then triple integrals for functions of three variables. We begin investigating in Preview Activity 11.7.

Preview Activity 11.7. Consider a solid piece granite in the shape of a box $B = \{(x, y, z) : 0 \leq x \leq 4, 0 \leq y \leq 6, 0 \leq z \leq 8\}$, whose density varies from point to point. Let $\delta(x, y, z)$ represent the mass density of the piece of granite at point (x, y, z) in kilograms per cubic meter (so we are measuring x, y, and z in meters). Our goal is to find the mass of this solid.

Recall that if the density was constant, we could find the mass by multiplying the density and volume; since the density varies from point to point, we will use the approach we did with two-variable lamina problems, and slice the solid into small pieces on which the density is roughly constant.

Partition the interval $[0, 4]$ into 2 subintervals of equal length, the interval $[0, 6]$ into 3 subintervals of equal length, and the interval $[0, 8]$ into 2 subintervals of equal length. This partitions the box B into sub-boxes as shown in Figure 11.33.

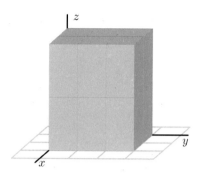

Figure 11.33: A partitioned three-dimensional domain.

(a) Let $0 = x_0 < x_1 < x_2 = 4$ be the endpoints of the subintervals of $[0, 4]$ after partitioning. Draw a picture of Figure 11.33 and label these endpoints on your drawing. Do likewise with $0 = y_0 < y_1 < y_2 < y_3 = 6$ and $0 = z_0 < z_1 < z_2 = 8$

What is the length Δx of each subinterval $[x_{i-1}, x_i]$ for i from 1 to 2? the length of Δy? of Δz?

(b) The partitions of the intervals $[0, 4]$, $[0, 6]$ and $[0, 8]$ partition the box B into sub-boxes. How many sub-boxes are there? What is volume ΔV of each sub-box?

(c) Let B_{ijk} denote the sub-box $[x_{i-1}, x_i] \times [y_{j-1}, y_j] \times [z_{k-1}, z_k]$. Say that we choose a point $(x_{ijk}^*, y_{ijk}^*, z_{ijk}^*)$ in the i, j, kth sub-box for each possible combination of i, j, k. What is the meaning of $\delta(x_{ijk}^*, y_{ijk}^*, z_{ijk}^*)$? What physical quantity will $\delta(x_{ijk}^*, y_{ijk}^*, z_{ijk}^*)\Delta V$ approximate?

(d) What final step(s) would it take to determine the exact mass of the piece of granite?

Triple Riemann Sums and Triple Integrals

Through the application of a mass density distribution over a three-dimensional solid, Preview Activity 11.7 suggests that the generalization from double Riemann sums of functions of two variables to triple Riemann sums of functions of three variables is natural. In the same way, so is the generalization from double integrals to triple integrals. By simply adding a z-coordinate to

11.7. TRIPLE INTEGRALS

our earlier work, we can define both a triple Riemann sum and the corresponding triple integral.

The Triple Riemann Sum.

Definition 11.3. Let $f = f(x, y, z)$ be a continuous function on a box $B = [a, b] \times [c, d] \times [r, s]$. The **triple Riemann sum of** f **over** B is created as follows.

- Partition the interval $[a, b]$ into m subintervals of equal length $\Delta x = \frac{b-a}{m}$. Let $x_0, x_1, \ldots, x_m$ be the endpoints of these subintervals, where $a = x_0 < x_1 < x_2 < \cdots < x_m = b$. Do likewise with the interval $[c, d]$ using n subintervals of equal length $\Delta y = \frac{d-c}{n}$ to generate $c = y_0 < y_1 < y_2 < \cdots < y_n = d$, and with the interval $[r, s]$ using ℓ subintervals of equal length $\Delta z = \frac{s-r}{\ell}$ to have $r = z_0 < z_1 < z_2 < \cdots < z_\ell = s$.

- Let B_{ijk} be the sub-box of B with opposite vertices $(x_{i-1}, y_{j-1}, z_{k-1})$ and (x_i, y_j, z_k) for i between 1 and m, j between 1 and n, and k between 1 and ℓ. The volume of each B_{ijk} is $\Delta V = \Delta x \cdot \Delta y \cdot \Delta z$.

- Let $(x_{ijk}^*, y_{ijk}^*, z_{ijk}^*)$ be a point in box B_{ijk} for each i, j, and k. The resulting triple Riemann sum for f on B is

$$\sum_{i=1}^{m} \sum_{j=1}^{n} \sum_{k=1}^{\ell} f(x_{ijk}^*, y_{ijk}^*, z_{ijk}^*) \cdot \Delta V.$$

If $f(x, y, z)$ represents the mass density of the box B, then, as we saw in Preview Activity 11.7, the triple Riemann sum approximates the total mass of the box B. In order to find the exact mass of the box, we need to let the number of sub-boxes increase without bound (in other words, let m, n, and ℓ go to infinity); in this case, the finite sum of the mass approximations becomes the actual mass of the solid B. More generally, we have the following definition of the triple integral.

The Triple Integral Over a Box.

Definition 11.4. With following notation defined as in a triple Riemann sum, the **triple integral of** f **over** B is

$$\iiint_B f(x, y, z)\, dV = \lim_{m,n,\ell \to \infty} \sum_{i=1}^{m} \sum_{j=1}^{n} \sum_{k=1}^{\ell} f(x_{ijk}^*, y_{ijk}^*, z_{ijk}^*) \cdot \Delta V.$$

As we noted earlier, if $f(x, y, z)$ represents the density of the solid B at each point (x, y, z), then

$$M = \iiint_B f(x, y, z)\, dV$$

is the mass of B. Even more importantly, for any continuous function f over the solid B, we can use a triple integral to determine the average value of f over B, $f_{\text{AVG}(B)}$. We note this generalization of our work with functions of two variables along with several others in the following important boxed information. Note that each of these quantities may actually be considered over a general domain S in $\mathbb{R}^3$, not simply a box, B.

- The triple integral
$$V(S) = \iiint_S 1\, dV$$
represents the **volume** of the solid S.

- The **average value** of the function $f = f(x, y, x)$ over a solid domain S is given by
$$f_{\text{AVG}(S)} = \left(\frac{1}{V(S)}\right) \iiint_S f(x, y, z)\, dV,$$
where $V(S)$ is the volume of the solid S.

- The **center of mass** of the solid S with density $\delta = \delta(x, y, z)$ is $(\overline{x}, \overline{y}, \overline{z})$, where
$$\overline{x} = \frac{\iiint_S x\, \delta(x, y, z)\, dV}{M}, \quad \overline{y} = \frac{\iiint_S y\, \delta(x, y, z)\, dV}{M}, \quad \overline{z} = \frac{\iiint_S z\, \delta(x, y, z)\, dV}{M},$$
and $M = \iiint_S \delta(x, y, z)\, dV$ is the mass of the solid S.

In the Cartesian coordinate system, the volume element dV is $dz\, dy\, dx$, and, as a consequence, a triple integral of a function f over a box $B = [a, b] \times [c, d] \times [r, s]$ in Cartesian coordinates can be evaluated as an iterated integral of the form
$$\iiint_B f(x, y, z)\, dV = \int_a^b \int_c^d \int_r^s f(x, y, z)\, dz\, dy\, dx.$$

If we want to evaluate a triple integral as an iterated integral over a solid S that is not a box, then we need to describe the solid in terms of variable limits.

Activity 11.20.

(a) Set up and evaluate the triple integral of $f(x, y, z) = x - y + 2z$ over the box $B = [-2, 3] \times [1, 4] \times [0, 2]$.

(b) Let S be the solid cone bounded by $z = \sqrt{x^2 + y^2}$ and $z = 3$. A picture of S is shown at right in Figure 11.34. Our goal in what follows is to set up an iterated integral of the form
$$\int_{x=?}^{x=?} \int_{y=?}^{y=?} \int_{z=?}^{z=?} \delta(x, y, z)\, dz\, dy\, dx \tag{11.5}$$
to represent the mass of S in the setting where $\delta(x, y, z)$ tells us the density of S at the point (x, y, z). Our particular task is to find the limits on each of the three integrals.

11.7. TRIPLE INTEGRALS

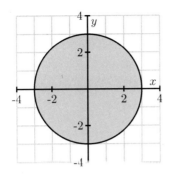 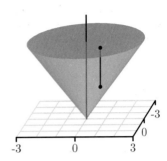

Figure 11.34: At right, the cone; at left, its projection.

i. If we think about slicing up the solid, we can consider slicing the domain of the solid's projection onto the xy-plane (just as we would slice a two-dimensional region in $\mathbb{R}^2$), and then slice in the z-direction as well. The projection of the solid is onto the xy-plane is shown at left in Figure 11.34. If we decide to first slice the domain of the solid's projection perpendicular to the x-axis, over what range of constant x-values would we have to slice?

ii. If we continue with slicing the domain, what are the limits on y on a typical slice? How do these depend on x? What, therefore, are the limits on the middle integral?

iii. Finally, now that we have thought about slicing up the two-dimensional domain that is the projection of the cone, what are the limits on z in the innermost integral? Note that over any point (x, y) in the plane, a vertical slice in the z direction will involve a range of values from the cone itself to its flat top. In particular, observe that at least one of these limits is not constant but depends on x and y.

iv. In conclusion, write an iterated integral of the form (11.5) that represents the mass of the cone S.

◁

Note well: When setting up iterated integrals, the limits on a given variable can be *only* in terms of the remaining variables. In addition, there are multiple different ways we can choose to set up such an integral. For example, two possibilities for iterated integrals that represent a triple integral $\iiint_S f(x, y, z) \, dV$ over a solid S are

- $\displaystyle\int_a^b \int_{g_1(x)}^{g_2(x)} \int_{h_1(x,y)}^{h_2(x,y)} f(x, y, z) \, dz \, dy \, dx$

- $\displaystyle\int_r^s \int_{p_1(z)}^{p_2(z)} \int_{q_1(x,z)}^{q_2(x,z)} f(x, y, z) \, dy \, dx \, dz$

where $g_1, g_2, h_1, h_2, p_1, p_2, q_1$, and q_2 are functions of the indicated variables. There are four other options beyond the two stated here, since the variables x, y, and z can (theoretically) be arranged in any order. Of course, in many circumstances, an insightful choice of variable order will make it easier to set up an iterated integral, just as was the case when we worked with double integrals.

Example 11.4. Find the mass of the tetrahedron in the first octant bounded by the coordinate planes and the plane $x + 2y + 3z = 6$ if the density at point (x, y, z) is given by $\delta(x, y, z) = x + y + z$. A picture of the solid tetrahedron is shown at left in Figure 11.35.

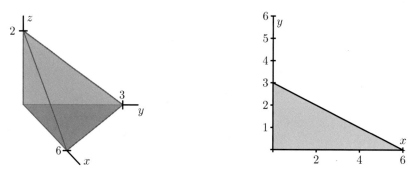

Figure 11.35: The tetrahedron and its projection.

We find the mass, M, of the tetrahedron by the triple integral

$$M = \iiint_S \delta(x, y, z) \, dV,$$

where S is the solid tetrahedron described above. In this example, we choose to integrate with respect to z first for the innermost integral. The top of the tetrahedron is given by the equation

$$x + 2y + 3z = 6;$$

solving for z then yields

$$z = \frac{1}{3}(6 - x - 2y).$$

The bottom of the tetrahedron is the xy-plane, so the limits on z in the iterated integral will be $0 \leq z \leq \frac{1}{3}(6 - x - 2y)$.

To find the bounds on x and y we project the tetrahedron onto the xy-plane; this corresponds to setting $z = 0$ in the equation $z = \frac{1}{3}(6 - x - 2y)$. The resulting relation between x and y is

$$x + 2y = 6.$$

The right image in Figure 11.35 shows the projection of the tetrahedron onto the xy-plane.

If we choose to integrate with respect to y for the middle integral in the iterated integral, then the lower limit on y is the x-axis and the upper limit is the hypotenuse of the triangle. Note that the hypotenuse joins the points $(6, 0)$ and $(0, 3)$ and so has equation $y = 3 - \frac{1}{2}x$. Thus, the bounds

11.7. TRIPLE INTEGRALS

on y are $0 \leq y \leq 3 - \frac{1}{2}x$. Finally, the x values run from 0 to 6, so the iterated integral that gives the mass of the tetrahedron is

$$M = \int_0^6 \int_0^{3-(1/2)x} \int_0^{(1/3)(6-x-2y)} x + y + z \, dz \, dy \, dx. \tag{11.6}$$

Evaluating the triple integral gives us

$$\begin{aligned} M &= \int_0^6 \int_0^{3-(1/2)x} \int_0^{(1/3)(6-x-2y)} x + y + z \, dz \, dy \, dx \\ &= \int_0^6 \int_0^{3-(1/2)x} \left[xz + yz + \frac{z^2}{2} \right]\bigg|_0^{(1/3)(6-x-2y)} dy \, dx \\ &= \int_0^6 \int_0^{3-(1/2)x} \frac{4}{3}x - \frac{5}{18}x^2 - \frac{2}{9}xy + \frac{4}{3}y - \frac{4}{9}y^2 + 2 \, dy \, dx \\ &= \int_0^6 \left[\frac{4}{3}xy - \frac{5}{18}x^2 y - \frac{7}{18}xy^2 + \frac{1}{3}y^2 - \frac{4}{27}y^3 + 2y \right]\bigg|_0^{3-(1/2)x} dx \\ &= \int_0^6 5 + \frac{1}{2}x - \frac{7}{12}x^2 + \frac{13}{216}x^3 \, dx \\ &= \left[5x + \frac{1}{4}x^2 - \frac{7}{36}x^3 + \frac{13}{864}x^4 \right]\bigg|_0^6 \\ &= \frac{33}{2}. \end{aligned}$$

Setting up limits on iterated integrals can require considerable geometric intuition. It is important to not only create carefully labeled figures, but also to think about how we wish to slice the solid. Further, note that when we say "we will integrate first with respect to x," by "first" we are referring to the innermost integral in the iterated integral. The next activity explores several different ways we might set up the integral in the preceding example.

Activity 11.21.

There are several other ways we could have set up the integral to give the mass of the tetrahedron in Example 11.4.

(a) How many different iterated integrals could be set up that are equal to the integral in Equation (11.6)?

(b) Set up an iterated integral, integrating first with respect to z, then x, then y that is equivalent to the integral in Equation (11.6). Before you write down the integral, think about Figure 11.35, and draw an appropriate two-dimensional image of an important projection.

(c) Set up an iterated integral, integrating first with respect to y, then z, then x that is equivalent to the integral in Equation (11.6). As in (b), think carefully about the geometry first.

(d) Set up an iterated integral, integrating first with respect to x, then y, then z that is equivalent to the integral in Equation (11.6).

◁

Now that we have begun to understand how to set up iterated triple integrals, we can apply them to determine important quantities, such as those found in the next activity.

Activity 11.22.

A solid S is bounded below by the square $z = 0$, $-1 \leq x \leq 1$, $-1 \leq y \leq 1$ and above by the surface $z = 2 - x^2 - y^2$. A picture of the solid is shown in Figure 11.36.

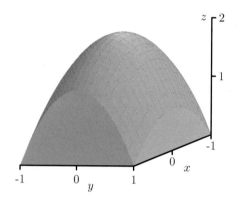

Figure 11.36: The solid bounded by the surface $z = 2 - x^2 - y^2$.

(a) Set up (but do not evaluate) an iterated integral to find the volume of the solid S.

(b) Set up (but do not evaluate) iterated integral expressions that will tell us the center of mass of S, if the density at point (x, y, z) is $\delta(x, y, z) = x^2 + 1$.

(c) Set up (but do not evaluate) an iterated integral to find the average density on S using the density function from part (b).

(d) Use technology appropriately to evaluate the iterated integrals you determined in (a), (b), and (c); does the location you determined for the center of mass make sense?

◁

Summary

- Let $f = f(x, y, z)$ be a continuous function on a box $B = [a, b] \times [c, d] \times [r, s]$. The triple integral of f over B is defined as

$$\iiint_B f(x, y, z)\, dV = \lim_{\Delta V \to 0} \sum_{i=1}^{m} \sum_{j=1}^{n} \sum_{k=1}^{l} f(x_{ijk}^*, y_{ijk}^*, z_{ijk}^*) \cdot \Delta V,$$

where the triple Riemann sum is defined in the usual way. The definition of the triple integral naturally extends to non-rectangular solid regions S.

11.7. TRIPLE INTEGRALS

- The triple integral $\iiint_S f(x,y,z)\,dV$ can tell us
  - the volume of the solid S if $f(x,y,z) = 1$,
  - the mass of the solid S if f represents the density of S at the point (x,y,z).

Moreover,
$$f_{\text{AVG}(S)} = \frac{1}{V(S)} \iiint_S f(x,y,z)\,dV,$$
is the average value of f over S.

Exercises

1. Consider the solid S that is bounded by the parabolic cylinder $y = x^2$ and the planes $z = 0$ and $z = 1 - y$ as shown in Figure 11.37.

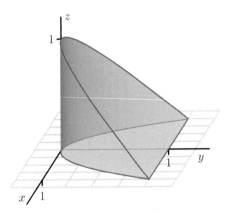

Figure 11.37: The solid bounded by $y = x^2$ and the planes $z = 0$ and $z = 1 - y$.

Assume the density of S is given by $\delta(x,y,z) = z$

(a) Set up (but do not evaluate) an iterated integral that represents the mass of S. Integrate first with respect to z, then y, then x. A picture of the projection of S onto the xy-plane is shown in Figure 11.38.

(b) Set up (but do not evaluate) an iterated integral that represents the mass of S. In this case, integrate first with respect to y, then z, then x. A picture of the projection of S onto the xz-plane is shown in Figure 11.39.

(c) Set up (but do not evaluate) an iterated integral that represents the mass of S. For this integral, integrate first with respect to x, then y, then z. A picture of the projection of S onto the yz-plane is shown in Figure 11.39.

(d) Which of these three orders of integration is the most natural to you? Why?

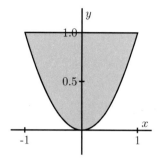 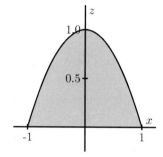 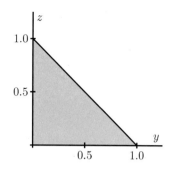

Figure 11.38: Projecting S onto the xy-plane.

Figure 11.39: Projecting S onto the xz-plane.

Figure 11.40: Projecting S onto the yz-plane.

2. This problem asks you to investigate the average value of some different quantities.

   (a) Set up, but do not evaluate, an iterated integral expression whose value is the average sum of all real numbers x, y, and z that have the following property: y is between 0 and 2, x is greater than or equal to 0 but cannot exceed $2y$, and z is greater than or equal to 0 but cannot exceed $x + y$.

   (b) Set up, but do not evaluate, an integral expression whose value represents the average value of $f(x, y, z) = x + y + z$ over the solid region in the first octant bounded by the surface $z = 4 - x - y^2$ and the coordinate planes $x = 0$, $y = 0$, $z = 0$.

   (c) How are the quantities in (a) and (b) similar? How are they different?

3. Consider the solid that lies between the paraboloids $z = g(x, y) = x^2 + y^2$ and $z = f(x, y) = 8 - 3x^2 - 3y^2$.

   (a) By eliminating the variable z, determine the curve of intersection between the two paraboloids, and sketch this curve in the x-y plane.

   (b) Set up, but do not evaluate, an iterated integral expression whose value determine the mass of the solid, integrating first with respect to x, then y, then z. Assume the the solid's density is given by $\delta(x, y, z) = \frac{1}{x^2+y^2+z^2+1}$.

   (c) Set up, but do not evaluate, iterated integral expressions whose values determine the mass of the solid using all possible remaining orders of integration. Use $\delta(x, y, z) = \frac{1}{x^2+y^2+z^2+1}$ as the density of the solid.

   (d) Set up, but do not evaluate, iterated integral expressions whose values determine the center of mass of the solid. Again, assume the the solid's density is given by $\delta(x, y, z) = \frac{1}{x^2+y^2+z^2+1}$.

   (e) Which coordinates of the center of mass can you determine *without* evaluating any integral expression? Why?

11.7. TRIPLE INTEGRALS

4. In each of the following problems, your task is to

    (i) sketch, by hand, the region over which you integrate

    (ii) set up iterated integral expressions which, when evaluated, will determine the value sought

    (iii) use appropriate technology to evaluate each iterated integral expression you develop

    Note well: in some problems you may be able to use a double rather than a triple integral, and polar coordinates may be helpful in some cases.

    (a) Consider the solid created by the region enclosed by the circular paraboloid $z = 4 - x^2 - y^2$ over the region R in the x-y plane enclosed by $y = -x$ and the circle $x^2 + y^2 = 4$ in the first, second, and fourth quadrants.

    Determine the solid's volume.

    (b) Consider the solid region that lies beneath the circular paraboloid $z = 9 - x^2 - y^2$ over the triangular region between $y = x$, $y = 2x$, and $y = 1$. Assuming that the solid has its density at point (x, y, z) given by $\delta(x, y, z) = xyz + 1$, measured in grams per cubic cm, determine the center of mass of the solid.

    (c) In a certain room in a house, the walls can be thought of as being formed by the lines $y = 0$, $y = 12 + x/4$, $x = 0$, and $x = 12$, where length is measured in feet. In addition, the ceiling of the room is vaulted and is determined by the plane $z = 16 - x/6 - y/3$. A heater is stationed in the corner of the room at $(0, 0, 0)$ and causes the temperature in the room at a particular time to be given by

    $$T(x, y, z) = \frac{80}{1 + \frac{x^2}{1000} + \frac{y^2}{1000} + \frac{z^2}{1000}}$$

    What is the average temperature in the room?

    (d) Consider the solid enclosed by the cylinder $x^2 + y^2 = 9$ and the planes $y + z = 5$ and $z = 1$. Assuming that the solid's density is given by $\delta(x, y, z) = \sqrt{x^2 + y^2}$, find the mass and center of mass of the solid.

11.8 Triple Integrals in Cylindrical and Spherical Coordinates

Motivating Questions

In this section, we strive to understand the ideas generated by the following important questions:

- What are the cylindrical coordinates of a point, and how are they related to Cartesian coordinates?

- What is the volume element in cylindrical coordinates? How does this inform us about evaluating a triple integral as an iterated integral in cylindrical coordinates?

- What are the spherical coordinates of a point, and how are they related to Cartesian coordinates?

- What is the volume element in spherical coordinates? How does this inform us about evaluating a triple integral as an iterated integral in spherical coordinates?

Introduction

We have encountered two different coordinate systems in $\mathbb{R}^2$ – the rectangular and polar coordinates systems – and seen how in certain situations, polar coordinates form a convenient alternative. In a similar way, there there turn out to be two additional natural coordinate systems in $\mathbb{R}^3$. Given that we are already familiar with the Cartesian coordinate system for $\mathbb{R}^3$, we next investigate the cylindrical and spherical coordinate systems (each of which builds upon polar coordinates in $\mathbb{R}^2$). In what follows, we will see how to convert among the different coordinate systems, how to evaluate triple integrals using them, and some situations in which these other coordinate systems prove advantageous.

Preview Activity 11.8. In the following questions, we investigate the two new coordinate systems that are the subject of this section: cylindrical and spherical coordinates. Our goal is to consider some examples of how to convert from rectangular coordinates to each of these systems, and vice versa. Triangles and trigonometry prove to be particularly important.

The cylindrical coordinates of a point in $\mathbb{R}^3$ are given by (r, θ, z) where r and θ are the polar coordinates of the point (x, y) and z is the same z coordinate as in Cartesian coordinates. An illustration is given in Figure 11.41.

(a) Find cylindrical coordinates for the point whose Cartesian coordinates are $(-1, \sqrt{3}, 3)$. Draw a labeled picture illustrating all of the coordinates.

(b) Find the Cartesian coordinates of the point whose cylindrical coordinates are $\left(2, \frac{5\pi}{4}, 1\right)$. Draw a labeled picture illustrating all of the coordinates.

The spherical coordinates of a point in $\mathbb{R}^3$ are ρ (rho), θ, and ϕ (phi), where ρ is the distance from the point to the origin, θ has the same interpretation it does in polar coordinates, and ϕ is the angle

11.8. TRIPLE INTEGRALS IN CYLINDRICAL AND SPHERICAL COORDINATES

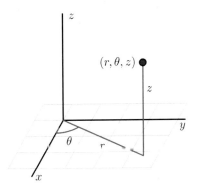

Figure 11.41: The cylindrical coordinates of a point.

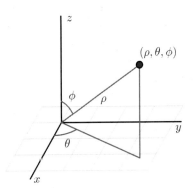

Figure 11.42: The spherical coordinates of a point.

between the positive z axis and the vector from the origin to the point, as illustrated in Figure 11.42.

For the following questions, consider the point P whose Cartesian coordinates are $(-2, 2, \sqrt{8})$.

(c) What is the distance from P to the origin? Your result is the value of ρ in the spherical coordinates of P.

(d) Determine the point that is the projection of P onto the xy-plane. Then, use this projection to find the value of θ in the polar coordinates of the projection of P that lies in the plane. Your result is also the value of θ for the spherical coordinates of the point.

(e) Based on the illustration in Figure 11.42, how is the angle ϕ determined by ρ and the z coordinate of P? Use a well-chosen right triangle to find the value of ϕ, which is the final component in the spherical coordinates of P. Draw a carefully labeled picture that clearly illustrates the values of ρ, θ, and ϕ in this example, along with the original rectangular coordinates of P.

(f) Based on your responses to (c), (d), and (e), if we are given the Cartesian coordinates (x, y, z) of a point Q, how are the values of ρ, θ, and ϕ in the spherical coordinates of Q determined by x, y, and z?

⋈

Cylindrical Coordinates

As we stated in Preview Activity 11.8, the cylindrical coordinates of a point are (r, θ, z), where r and θ are the polar coordinates of the point (x, y), and z is the same z coordinate as in Cartesian coordinates. The general situation is illustrated Figure 11.41.

11.8. TRIPLE INTEGRALS IN CYLINDRICAL AND SPHERICAL COORDINATES

Since we already know how to convert between rectangular and polar coordinates in the plane, and the z coordinate is identical in both Cartesian and cylindrical coordinates, the conversion equations between the two systems in $\mathbb{R}^3$ are essentially those we found for polar coordinates:

$$x = r\cos(\theta) \qquad y = r\sin(\theta) \qquad z = z$$
$$r^2 = x^2 + y^2 \qquad \tan(\theta) = \frac{y}{x} \qquad z = z.$$

Just as with rectangular coordinates, where we usually write z as a function of x and y to plot the resulting surface, in cylindrical coordinates, we often express z as a function of r and θ. In the following activity, we explore several basic equations in cylindrical coordinates and the corresponding surface each generates.

Activity 11.23.

In this activity, we graph some surfaces using cylindrical coordinates. To improve your intuition and test your understanding, you should first think about what each graph should look like before you plot it using technology.[8]

(a) Plot the graph of the cylindrical equation $r = 2$, where we restrict the values of θ and z to the intervals $0 \leq \theta \leq 2\pi$ and $0 \leq z \leq 2$. What familiar shape does the resulting surface take? How does this example suggest that we call these coordinates *cylindrical coordinates*?

(b) Plot the graph of the cylindrical equation $\theta = 2$, where we restrict the other variables to the values $0 \leq r \leq 2$ and $0 \leq z \leq 2$. What familiar surface results?

(c) Plot the graph of the cylindrical equation $z = 2$, using $0 \leq \theta \leq 2\pi$ and $0 \leq r \leq 2$. What does this surface look like?

(d) Plot the graph of the cylindrical equation $z = r$, where $0 \leq \theta \leq 2\pi$ and $0 \leq r \leq 2$. What familiar surface results?

(e) Plot the graph of the cylindrical equation $z = \theta$ for $0 \leq \theta \leq 4\pi$. What does this surface look like?

◁

As the name and Activity 11.23 suggest, cylindrical coordinates are useful for describing surfaces that are cylindrical in nature.

Triple Integrals in Cylindrical Coordinates

To evaluate a triple integral $\iiint_S f(x, y, z)\,dV$ as an iterated integral in Cartesian coordinates, we use the fact that the volume element dV is equal to $dz\,dy\,dx$ (which corresponds to the volume of a

---

[8]e.g., http://www.math.uri.edu/~bkaskosz/flashmo/cylin/ – to plot $r = 2$, set r to 2, θ to s, and z to t – to plot $\theta = \pi/3$, set $\theta = \pi/3$, $r = s$, and $z = t$, for example. Thanks to Barbara Kaskosz of URI and the Flash and Math team.

11.8. TRIPLE INTEGRALS IN CYLINDRICAL AND SPHERICAL COORDINATES

small box). To evaluate a triple integral in cylindrical coordinates, we similarly must understand the volume element dV in cylindrical coordinates.

Activity 11.24.

A picture of a cylindrical box, $B = \{(r, \theta, z) : r_1 \leq r \leq r_2, \theta_1 \leq \theta \leq \theta_2, z_1 \leq z \leq z_2\}$, is shown in Figure 11.43. Let $\Delta r = r_2 - r_1$, $\Delta \theta = \theta_2 - \theta_1$, and $\Delta z = z_2 - z_1$. We want to determine the volume ΔV of B in terms of $\Delta r, \Delta \theta, \Delta z, r, \theta$, and z.

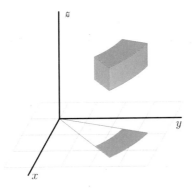

Figure 11.43: A cylindrical box.

(a) Appropriately label Δr, $\Delta \theta$, and Δz in Figure 11.43.

(b) Let ΔA be the area of the projection of the box, B, onto the xy-plane, which is shaded blue in Figure 11.43. Recall that we previously determined the area ΔA in polar coordinates in terms of r, Δr, and $\Delta \theta$. In light of the fact that we know ΔA and that z is the standard z coordinate from Cartesian coordinates, what is the volume ΔV in cylindrical coordinates?

◁

Activity 11.24 demonstrates that the volume element dV in cylindrical coordinates is given by $dV = r\, dz\, dr\, d\theta$, and hence the following rule holds in general.

Given a continuous function $f = f(x, y, z)$ over a region S in $\mathbb{R}^3$,

$$\iiint_S f(x, y, z)\, dV = \iiint_S f(r\cos(\theta), r\sin(\theta), z)\, r\, dz\, dr\, d\theta.$$

The latter expression is an **iterated integral in cylindrical coordinates**.

Of course, to complete the task of writing an iterated integral in cylindrical coordinates, we need to determine the limits on the three integrals: θ, r, and z. In the following activity, we explore how to do this in several situations where cylindrical coordinates are natural and advantageous.

Activity 11.25.

In each of the following questions, set up, but do not evaluate, the requested integral expression.

(a) Let S be the solid bounded above by the graph of $z = x^2 + y^2$ and below by $z = 0$ on the unit circle. Determine an iterated integral expression in cylindrical coordinates that gives the volume of S.

(b) Suppose the density of the cone defined by $r = 1 - z$, with $z \geq 0$, is given by $\delta(r, \theta, z) = z$. A picture of the cone is shown in Figure 11.44, and the projection of the cone onto the xy-plane in given in Figure 11.45. Set up an iterated integral in cylindrical coordinates that gives the mass of the cone.

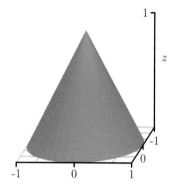

Figure 11.44: The cylindrical cone $r = 1 - z$.

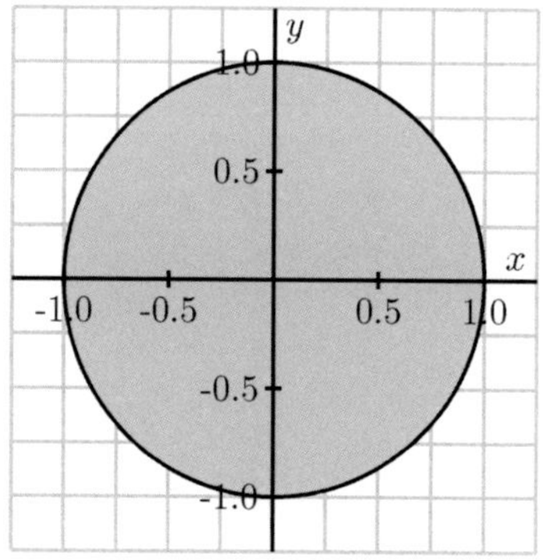

Figure 11.45: The projection into the xy-plane.

(c) Determine an iterated integral expression in cylindrical coordinates whose value is the volume of the solid bounded below by the cone $z = \sqrt{x^2 + y^2}$ and above by the cone $z = 4 - \sqrt{x^2 + y^2}$. A picture is shown in Figure 11.46.

◁

Spherical Coordinates

As we saw in Preview Activity 11.8, the spherical coordinates of a point in 3-space have the form (ρ, θ, ϕ), where ρ is the distance from the point to the origin, θ has the same meaning as in polar coordinates, and ϕ is the angle between the positive z axis and the vector from the origin to the point. The overall situation is illustrated in Figure 11.42.

The example in Preview Activity 11.8 suggests that given a point in rectangular coordinates, (x, y, z), we can find the corresponding spherical coordinates. Indeed, this holds generally by

11.8. TRIPLE INTEGRALS IN CYLINDRICAL AND SPHERICAL COORDINATES

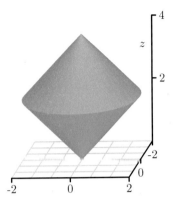

Figure 11.46: A solid bounded by the cones $z = \sqrt{x^2 + y^2}$ and $z = 4 - \sqrt{x^2 + y^2}$.

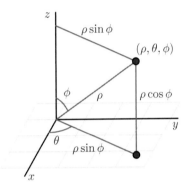

Figure 11.47: Converting from spherical to Cartesian coordinates.

using the relationships

$$\rho = \sqrt{x^2 + y^2 + z^2} \qquad \tan(\theta) = \frac{y}{x} \qquad \cos(\phi) = \frac{z}{\rho}$$

where in the latter two equations, we require $x \neq 0$ and $\rho \neq 0$.

To convert from given spherical coordinates to Cartesian coordinates, we use the equations

$$x = \rho \sin(\phi) \cos(\theta) \qquad y = \rho \sin(\phi) \sin(\theta) \qquad z = \rho \cos(\phi),$$

as illustrated in Figure 11.47.

When it comes to thinking about particular surfaces in spherical coordinates, similar to our work with cylindrical and Cartesian coordinates, we usually write ρ as a function of θ and ϕ; this is a natural analog to polar coordinates, where we often think of our distance from the origin in the plane as being a function of θ. In spherical coordinates, we likewise often view ρ as a function of θ and ϕ, thus viewing distance from the origin as a function of two key angles.

In the following activity, we explore several basic equations in spherical coordinates and the surfaces they generate.

Activity 11.26.

In this activity, we graph some surfaces using spherical coordinates. To improve your intuition and test your understanding, you should first think about what each graph should look like before you plot it using technology.[9]

(a) Plot the graph of $\rho = 1$, where θ and ϕ are restricted to the intervals $0 \leq \theta \leq 2\pi$ and $0 \leq \phi \leq \pi$. What is the resulting surface? How does this particular example demonstrate the reason for the name of this coordinate system?

(b) Plot the graph of $\phi = \frac{\pi}{3}$, where ρ and θ are restricted to the intervals $0 \leq \rho \leq 1$ and $0 \leq \theta \leq 2\pi$. What familiar surface results?

(c) Plot the graph of $\theta = \frac{\pi}{6}$, for $0 \leq \rho \leq 1$ and $0 \leq \phi \leq \pi$. What familiar shape arises?

(d) Plot the graph of $\rho = \theta$, for $0 \leq \phi \leq \pi$ and $0 \leq \theta \leq 2\pi$. How does the resulting surface appear?

◁

As the name and Activity 11.26 indicate, spherical coordinates are particularly useful for describing surfaces that are spherical in nature; they are also convenient for working with certain conical surfaces.

Triple Integrals in Spherical Coordinates

As with rectangular and cylindrical coordinates, a triple integral $\iiint_S f(x, y, z)\, dV$ in spherical coordinates can be evaluated as an iterated integral once we understand the volume element dV.

Activity 11.27.

To find the volume element dV in spherical coordinates, we need to understand how to determine the volume of a spherical box of the form $\rho_1 \leq \rho \leq \rho_2$ (with $\Delta\rho = \rho_2 - \rho_1$), $\phi_1 \leq \phi \leq \phi_2$ (with $\Delta\phi = \phi_2 - \phi_1$), and $\theta_1 \leq \theta \leq \theta_2$ (with $\Delta\theta = \theta_2 - \theta_1$). An illustration of such a box is given in Figure 11.48. This spherical box is a bit more complicated than the cylindrical box we encountered earlier. In this situation, it is easier to approximate the volume ΔV than to compute it directly. Here we can approximate the volume ΔV of this spherical box with the volume of a Cartesian box whose sides have the lengths of the sides of this spherical box. In other words,

$$\Delta V \approx |PS|\, |\widehat{PR}|\, |\widehat{PQ}|,$$

where $|\widehat{PR}|$ denotes the length of the circular arc from P to R.

---

[9] e.g., http://www.flashandmath.com/mathlets/multicalc/paramsphere/surf_graph_sphere.html – to plot $\rho = 2$, set ρ to 2, θ to s, and ϕ to t, for example. Thanks to Barbara Kaskosz of URI and the Flash and Math team.

11.8. TRIPLE INTEGRALS IN CYLINDRICAL AND SPHERICAL COORDINATES

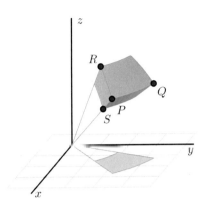

Figure 11.48: A spherical box.

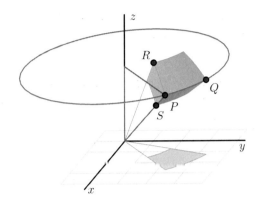

Figure 11.49: A spherical volume element.

(a) What is the length $|PS|$ in terms of ρ?

(b) What is the length of the arc $\widehat{PR}$? (Hint: The arc $\widehat{PR}$ is an arc of a circle of radius ρ_2, and arc length along a circle is the product of the angle measure (in radians) and the circle's radius.)

(c) What is the length of the arc $\widehat{PQ}$? (Hint: The arc $\widehat{PQ}$ lies on a horizontal circle as illustrated in Figure 11.49. What is the radius of this circle?)

(d) Use your work in (a), (b), and (c) to determine an approximation for ΔV in spherical coordinates.

◁

Letting $\Delta\rho$, $\Delta\phi$ and $\Delta\theta$ go to 0, it follows from the final result in Activity 11.27 that $dV = \rho^2 \sin(\phi)\, d\rho\, d\phi\, d\theta$ in spherical coordinates, and thus allows us to state the following general rule.

Given a continuous function $f = f(x, y, z)$ over a region S in $\mathbb{R}^3$,

$$\iiint_S f(x,y,z)\, dV = \iiint_S f(\rho\sin(\phi)\cos(\theta), \rho\sin(\phi)\sin(\theta), \rho\cos(\phi))\, \rho^2 \sin(\phi)\, d\rho\, d\phi\, d\theta.$$

The latter expression is an **iterated integral in spherical coordinates**.

Finally, in order to actually evaluate an iterated integral in spherical coordinates, we must of course determine the limits of integration in θ, ϕ, and ρ. The process is similar to our earlier work in the other two coordinate systems.

Activity 11.28.

We can use spherical coordinates to help us more easily understand some natural geometric objects.

(a) Recall that the sphere of radius a has spherical equation $\rho = a$. Set up and evaluate an iterated integral in spherical coordinates to determine the volume of a sphere of radius a.

(b) Set up, but do not evaluate, an iterated integral expression in spherical coordinates whose value is the mass of the solid obtained by removing the cone $\phi = \frac{\pi}{4}$ from the sphere $\rho = 2$ if the density δ at the point (x, y, z) is $\delta(x, y, z) = \sqrt{x^2 + y^2 + z^2}$. An illustration of the solid is shown in Figure 11.50.

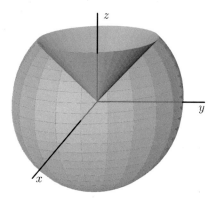

Figure 11.50: The solid cut from the sphere $\rho = 2$ by the cone $\phi = \frac{\pi}{4}$.

◁

Summary

- The cylindrical coordinates of a point P are (r, θ, z) where r is the distance from the origin to the projection of P onto the xy-plane, θ is the angle that the projection of P onto the xy-plane makes with the positive x-axis, and z is the vertical distance from P to the projection of P onto the xy-plane. When P has rectangular coordinates (x, y, z), it follows that its cylindrical coordinates are given by

$$r^2 = x^2 + y^2, \quad \tan(\theta) = \frac{y}{x}, \quad z = z.$$

When P has given cylindrical coordinates (r, θ, z), its rectangular coordinates are

$$x = r\cos(\theta), \quad y = r\sin(\theta), \quad z = z.$$

- The volume element dV in cylindrical coordinates is $dV = r\, dz\, dr\, d\theta$. Hence, a triple integral $\iiint_S f(x, y, z)\, dA$ can be evaluated as the iterated integral

$$\iiint_S f(r\cos(\theta), r\sin(\theta), z)\, r\, dz\, dr\, d\theta.$$

11.8. TRIPLE INTEGRALS IN CYLINDRICAL AND SPHERICAL COORDINATES

- The spherical coordinates of a point P in 3-space are ρ (rho), ϕ (phi), and θ, where ρ is the distance from P to the origin, ϕ is the angle between the positive z axis and the vector from the origin to P, and θ is the angle that the projection of P onto the xy-plane makes with the positive x-axis. When P has Cartesian coordinates (x, y, z), the spherical coordinates are given by

$$\rho^2 = x^2 + y^2 + z^2, \qquad \tan(\theta) = \frac{y}{x}, \qquad \cos(\phi) = \frac{z}{\rho}.$$

Given the point P in spherical coordinates (ρ, ϕ, θ), its rectangular coordinates are

$$x = \rho \sin(\phi) \cos(\theta), \qquad y = \rho \sin(\phi) \sin(\theta), \qquad z = \rho \cos(\phi).$$

- The volume element dV in spherical coordinates is $dV = \rho^2 \sin(\phi)\, d\rho\, d\phi\, d\theta$. Thus, a triple integral $\iiint_S f(x, y, z)\, dA$ can be evaluated as the iterated integral

$$\iiint_S f(\rho \sin(\phi) \cos(\theta), \rho \sin(\phi) \sin(\theta), \rho \cos(\phi))\, \rho^2 \sin(\phi)\, d\rho\, d\phi\, d\theta.$$

Exercises

1. In each of the following questions, set up an iterated integral expression whose value determines the desired result. Then, evaluate the integral first by hand, and then using appropriate technology.

    (a) Find the volume of the "cap" cut from the solid sphere $x^2 + y^2 + z^2 = 4$ by the plane $z = 1$, as well as the z-coordinate of its centroid.

    (b) Find the x-coordinate of the center of mass of the portion of the unit sphere that lies in the first octant (i.e., where x, y, and z are all nonnegative). Assume that the density of the solid given by $\delta(x, y, z) = \frac{1}{1+x^2+y^2+z^2}$.

    (c) Find the volume of the solid bounded below by the x-y plane, on the sides by the sphere $\rho = 2$, and above by the cone $\phi = \pi/3$.

    (d) Find the z coordinate of the center of mass of the region that is bounded above by the surface $z = \sqrt{\sqrt{x^2+y^2}}$, on the sides by the cylinder $x^2 + y^2 = 4$, and below by the x-y plane. Assume that the density of the solid is uniform and constant.

    (e) Find the volume of the solid that lies outside the sphere $x^2 + y^2 + z^2 = 1$ and inside the sphere $x^2 + y^2 + z^2 = 2z$.

2. For each of the following questions, (a) sketch the region of integration, (b) change the coordinate system in which the iterated integral is written to one of the remaining two, and (c) evaluate the iterated integral you deem easiest to evaluate by hand.

    (a) $\displaystyle\int_0^1 \int_0^{\sqrt{1-x^2}} \int_{\sqrt{x^2+y^2}}^{\sqrt{2-x^2-y^2}} xy\, dz\, dy\, dx$

(b) $\int_0^{\pi/2} \int_0^{\pi} \int_0^1 \rho^2 \sin(\phi)\, d\rho\, d\phi\, d\theta$

(c) $\int_0^{2\pi} \int_0^1 \int_r^1 r^2 \cos(\theta)\, dz\, dr\, d\theta$

3. Consider the solid region S bounded above by the paraboloid $z = 16 - x^2 - y^2$ and below by the paraboloid $z = 3x^2 + 3y^2$.

   (a) Describe parametrically the curve in $\mathbb{R}^3$ in which these two surfaces intersect.

   (b) In terms of x and y, write an equation to describe the projection of the curve onto the x-y plane.

   (c) What coordinate system do you think is most natural for an iterated integral that gives the volume of the solid?

   (d) Set up, but do not evaluate, an iterated integral expression whose value is average z-value of points in the solid region S.

   (e) Use technology to plot the two surfaces and evaluate the integral in (c). Write at least one sentence to discuss how your computations align with your intuition about where the average z-value of the solid should fall.

11.9 Change of Variables

> **Motivating Questions**
>
> *In this section, we strive to understand the ideas generated by the following important questions:*
>
> - What is a change of variables?
> - What is the Jacobian, and how is it related to a change of variables?

Introduction

In single variable calculus, we encountered the idea of a change of variable in a definite integral through the method of substitution. For example, given the definite integral

$$\int_0^2 2x(x^2+1)^3 \, dx,$$

we naturally consider the change of variable $u = x^2 + 1$. From this substitution, it follows that $du = 2x \, dx$, and since $x = 0$ implies $u = 1$ and $x = 2$ implies $u = 5$, we have transformed the original integral in x into a new integral in u. In particular,

$$\int_0^2 2x(x^2+1)^3 \, dx = \int_1^5 u^3 \, du.$$

The latter integral, of course, is far easier to evaluate.

Through our work with polar, cylindrical, and spherical coordinates, we have already implicitly seen some of the issues that arise in using a change of variables with two or three variables present. In what follows, we seek to understand the general ideas behind any change of variables in a multiple integral.

Preview Activity 11.9. Consider the double integral

$$I = \iint_D x^2 + y^2 \, dA, \tag{11.7}$$

where D is the upper half of the unit disk.

(a) Write the double integral I given in Equation (11.7) as an iterated integral in rectangular coordinates.

(b) Write the double integral I given in Equation (11.7) as an iterated integral in polar coordinates.

When we write the double integral (11.7) as an iterated integral in polar coordinates we make a change of variables, namely

$$x = r\cos(\theta) \quad \text{and} \quad y = r\sin(\theta). \tag{11.8}$$

We also then have to change dA to $r\,dr\,d\theta$. This process also identifies a "polar rectangle" $[r_1, r_2] \times [\theta_1, \theta_2]$ with the original Cartesian rectangle, under the transformation[10] in Equation (11.8). The vertices of the polar rectangle are transformed into the vertices of a closed and bounded region in rectangular coordinates.

To work with a numerical example, let's now consider the polar rectangle P given by $[1, 2] \times [\frac{\pi}{6}, \frac{\pi}{4}]$, so that $r_1 = 1$, $r_2 = 2$, $\theta_1 = \frac{\pi}{6}$, and $\theta_2 = \frac{\pi}{4}$.

(c) Use the transformation determined by the equations in (11.8) to find the rectangular vertices that correspond to the polar vertices in the polar rectangle P. In other words, by substituting appropriate values of r and θ into the two equations in (11.8), find the values of the corresponding x and y coordinates for the vertices of the polar rectangle P. Label the point that corresponds to the polar vertex (r_1, θ_1) as (x_1, y_1), the point corresponding to the polar vertex (r_2, θ_1) as (x_2, y_2), the point corresponding to the polar vertex (r_1, θ_2) as (x_3, y_3), and the point corresponding to the polar vertex (r_2, θ_2) as (x_4, y_4).

(d) Draw a picture of the figure in rectangular coordinates that has the points (x_1, y_1), (x_2, y_2), (x_3, y_3), and (x_4, y_4) as vertices. (Note carefully that because of the trigonometric functions in the transformation, this region will not look like a Cartesian rectangle.) What is the area of this region in rectangular coordinates? How does this area compare to the area of the original polar rectangle?

⋈

Change of Variables in Polar Coordinates

The general idea behind a change of variables is suggested by Preview Activity 11.9. There, we saw that in a change of variables from rectangular coordinates to polar coordinates, a polar rectangle $[r_1, r_2] \times [\theta_1, \theta_2]$ gets mapped to a Cartesian rectangle under the transformation

$$x = r\cos(\theta) \quad \text{and} \quad y = r\sin(\theta).$$

The vertices of the polar rectangle P are transformed into the vertices of a closed and bounded region P' in rectangular coordinates. If we view the standard coordinate system as having the horizontal axis represent r and the vertical axis represent θ, then the polar rectangle P appears to

---

[10] A *transformation* is another name for function: here, the equations $x = r\cos(\theta)$ and $y = r\sin(\theta)$ define a function T by $T(r, \theta) = (r\cos(\theta), r\sin(\theta))$ so that T is a function (transformation) from $\mathbb{R}^2$ to $\mathbb{R}^2$. We view this transformation as mapping a version of the x-y plane where the axes are viewed as representing r and θ (the r-θ plane) to the familiar x-y plane.

11.9. CHANGE OF VARIABLES

us at left in Figure 11.51. The image P' of the polar rectangle P under the transformation given by (11.8) is shown at right in Figure 11.51. We thus see that there is a correspondence between a simple region (a traditional, right-angled rectangle) and a more complicated region (a fraction of an annulus) under the function T given by $T(r, \theta) = (r\cos(\theta), r\sin(\theta))$.

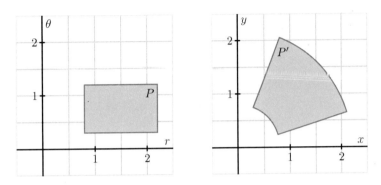

Figure 11.51: A rectangle P and its image P'.

Furthermore, as Preview Activity 11.9 suggest, it follows generally that for an original polar rectangle $P = [r_1, r_2] \times [\theta_1, \theta_2]$, the area of the transformed rectangle P' is given by $\frac{r_2+r_1}{2}\Delta r \Delta \theta$. Therefore, as Δr and $\Delta \theta$ go to 0 this area becomes the familiar area element $dA = r\, dr\, d\theta$ in polar coordinates. When we proceed to working with other transformations for different changes in coordinates, we have to understand how the transformation affects area so that we may use the correct area element in the new system of variables.

General Change of Coordinates

We first focus on double integrals. As with single integrals, we may be able to simplify a double integral of the form

$$\iint_D f(x,y)\, dA$$

by making a change of variables (that is, a substitution) of the form

$$x = x(s,t) \quad \text{and} \quad y = y(s,t)$$

where x and y are functions of new variables s and t. This transformation introduces a correspondence between a problem in the xy-plane and one in the the st-plane. The equations $x = x(s, t)$ and $y = y(s, t)$ convert s and t to x and y; we call these formulas the *change of variable* formulas. To complete the change to the new s, t variables, we need to understand the area element, dA, in this new system. The following activity helps to illustrate the idea.

Activity 11.29.

Consider the change of variables

$$x = s + 2t \quad \text{and} \quad y = 2s + \sqrt{t}.$$

Let's see what happens to the rectangle $T = [0,1] \times [1,4]$ in the st-plane under this change of variable.

(a) Draw a labeled picture of T in the st-plane.

(b) Find the image of the st-vertex $(0,1)$ in the xy-plane. Likewise, find the respective images of the other three vertices of the rectangle T: $(0,4)$, $(1,1)$, and $(1,4)$.

(c) In the xy-plane, draw a labeled picture of the image, T', of the original st-rectangle T. What appears to be the shape of the image, T'?

(d) To transform an integral with a change of variables, we need to determine the area element dA for image of the transformed rectangle. How would we find the area of the xy-figure T'? (Hint: Remember what the cross product of two vectors tells us.)

◁

Activity 11.29 presents the general idea of how a change of variables works. We partition a rectangular domain in the st system into subrectangles. Let $T = [a,b] \times [a + \Delta s, b + \Delta t]$ be one of these subrectangles. Then we transform this into a region T' in the standard xy Cartesian coordinate system. The region T' is called the *image* of T; the region T is the *pre-image* of T'. Although the sides of this xy region T' aren't necessarily straight (linear), we will approximate the element of area dA for this region with the area of the parallelogram whose sides are given by the vectors **v** and **w**, where **v** is the vector from $(x(a,b), y(a,b))$ to $(x(a+\Delta s, b), y(a+\Delta s, b))$, and **w** is the vector from $(x(a,b), y(a,b))$ to $(x(a, b+\Delta t), y(a, b+\Delta t))$.

An example of an image T' in the xy plane that results from a transformation of a rectangle T in the st plane is shown in Figure 11.52.

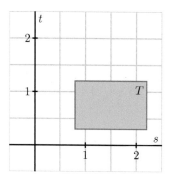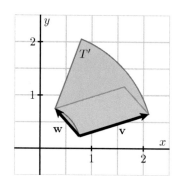

Figure 11.52: Approximating an area of an image resulting from a transformation.

The components of the vector **v** are
$$\mathbf{v} = \langle x(a+\Delta s, b) - x(a,b), y(a+\Delta s, b) - y(a,b), 0 \rangle$$
and similarly those for **w** are
$$\mathbf{w} = \langle x(a, b+\Delta t) - x(a,b), y(a, b+\Delta s) - y(a,b), 0 \rangle.$$

11.9. CHANGE OF VARIABLES

Slightly rewriting **v** and **w**, we have

$$\mathbf{v} = \left\langle \frac{x(a+\Delta s, b) - x(a,b)}{\Delta s}, \frac{y(a+\Delta s, b) - y(a,b)}{\Delta s}, 0 \right\rangle \Delta s, \text{ and}$$

$$\mathbf{w} = \left\langle \frac{x(a, b+\Delta t) - x(a,b)}{\Delta t}, \frac{y(a, b+\Delta s) - y(a,b)}{\Delta t}, 0 \right\rangle \Delta t.$$

For small Δs and Δt, the definition of the partial derivative tells us that

$$\mathbf{v} \approx \left\langle \frac{\partial x}{\partial s}(a,b), \frac{\partial y}{\partial s}(a,b), 0 \right\rangle \Delta s \quad \text{and} \quad \mathbf{w} \approx \left\langle \frac{\partial x}{\partial t}(a,b), \frac{\partial y}{\partial t}(a,b), 0 \right\rangle \Delta t.$$

Recall that the area of the parallelogram with sides **v** and **w** is the length of the cross product of the two vectors, $|\mathbf{v} \times \mathbf{w}|$. From this, we observe that

$$\mathbf{v} \times \mathbf{w} \approx \left\langle \frac{\partial x}{\partial s}(a,b), \frac{\partial y}{\partial s}(a,b), 0 \right\rangle \Delta s \times \left\langle \frac{\partial x}{\partial t}(a,b), \frac{\partial y}{\partial t}(a,b), 0 \right\rangle \Delta t$$

$$= \left\langle 0, 0, \frac{\partial x}{\partial s}(a,b)\frac{\partial y}{\partial t}(a,b) - \frac{\partial x}{\partial t}(a,b)\frac{\partial y}{\partial s}(a,b) \right\rangle \Delta s\, \Delta t.$$

Finally, by computing the magnitude of the cross product, we see that

$$|\mathbf{v} \times \mathbf{w}| \approx \left| \left\langle 0, 0, \frac{\partial x}{\partial s}(a,b)\frac{\partial y}{\partial t}(a,b) - \frac{\partial x}{\partial t}(a,b)\frac{\partial y}{\partial s}(a,b) \right\rangle \Delta s\, \Delta t \right|$$

$$= \left| \frac{\partial x}{\partial s}(a,b)\frac{\partial y}{\partial t}(a,b) - \frac{\partial x}{\partial t}(a,b)\frac{\partial y}{\partial s}(a,b) \right| \Delta s\, \Delta t.$$

Therefore, as the number of subdivisions increases without bound in each direction, Δs and Δt both go to zero, and we have

$$dA = \left| \frac{\partial x}{\partial s}\frac{\partial y}{\partial t} - \frac{\partial x}{\partial t}\frac{\partial y}{\partial s} \right| ds\, dt. \tag{11.9}$$

Equation (11.9) hence determines the general change of variable formula in a double integral, and we can now say that

$$\iint_T f(x,y)\, dA = \iint_R f(x,y)\, dy\, dx = \iint_{T'} f(x(s,t), y(s,t)) \left| \frac{\partial x}{\partial s}\frac{\partial y}{\partial t} - \frac{\partial x}{\partial t}\frac{\partial y}{\partial s} \right| ds\, dt.$$

The quantity

$$\left| \frac{\partial x}{\partial s}\frac{\partial y}{\partial t} - \frac{\partial x}{\partial t}\frac{\partial y}{\partial s} \right|$$

is called the *Jacobian*, and we denote the Jacobian using the shorthand notation

$$\left|\frac{\partial(x,y)}{\partial(s,t)}\right| = \left|\frac{\partial x}{\partial s}\frac{\partial y}{\partial t} - \frac{\partial x}{\partial t}\frac{\partial y}{\partial s}\right|.^{11}$$

To summarize, the preceding change of variable formula that we have derived now follows.

Change of Variables in a Double Integral. Suppose a change of variables $x = x(s,t)$ and $y = y(s,t)$ transforms a closed and bounded region R in the st-plane into a closed and bounded region R' in the xy-plane. Under modest conditions (that are studied in advanced calculus), it follows that

$$\iint_{R'} f(x,y)\,dA = \iint_R f(x(s,t), y(s,t)) \left|\frac{\partial(x,y)}{\partial(s,t)}\right| ds\,dt.$$

Activity 11.30.

Find the Jacobian when changing from rectangular to polar coordinates. That is, for the transformation given by $x = r\cos(\theta)$, $y = r\sin(\theta)$, determine a simplified expression for the quantity

$$\left|\frac{\partial x}{\partial r}\frac{\partial y}{\partial \theta} - \frac{\partial x}{\partial \theta}\frac{\partial y}{\partial r}\right|.$$

What do you observe about your result? How is this connected to our earlier work with double integrals in polar coordinates?

◁

Given a particular double integral, it is natural to ask, "how can we find a useful change of variables?" There are two general factors to consider: if the integrand is particularly difficult, we might choose a change of variables that would make the integrand easier; or, given a complicated region of integration, we might choose a change of variables that transforms the region of integration into one that has a simpler form. These ideas are illustrated in the next activities.

Activity 11.31.

Consider the problem of finding the area of the region D' defined by the ellipse $x^2 + \frac{y^2}{4} = 1$. Here we will make a change of variables so that the pre-image of the domain is a circle.

(a) Let $x(s,t) = s$ and $y(s,t) = 2t$. Explain why the pre-image of the original ellipse (which lies in the xy plane) is the circle $s^2 + t^2 = 1$ in the st-plane.

---

[11] If you are familiar with determinants of matrices, we can can also represent $\frac{\partial(x,y)}{\partial(s,t)}$ as the determinant of a 2×2 matrix

$$\frac{\partial(x,y)}{\partial(s,t)} = \begin{vmatrix} \frac{\partial x}{\partial s} & \frac{\partial x}{\partial t} \\ \frac{\partial y}{\partial s} & \frac{\partial y}{\partial t} \end{vmatrix}.$$

11.9. CHANGE OF VARIABLES

(b) Recall that the area of the ellipse D' is determined by the double integral $\iint_{D'} 1\, dA$. Explain why

$$\iint_{D'} 1\, dA = \iint_D 2\, ds\, dt$$

where D is the disk bounded by the circle $s^2 + t^2 = 1$. In particular, explain the source of the "2" in the st integral.

(c) Without evaluating any of the integrals present, explain why the area of the original elliptical region D' is 2π.

◁

Activity 11.32.

Let D' be the region in the xy-plane bounded by the lines $y = 0$, $x = 0$, and $x + y = 1$. We will evaluate the double integral

$$\iint_{D'} \sqrt{x+y}(x-y)^2\, dA \tag{11.10}$$

with a change of variables.

(a) Sketch the region D' in the xy plane.

(b) We would like to make a substitution that makes the integrand easier to antidifferentiate. Let $s = x+y$ and $t = x-y$. Explain why this should make antidifferentiation easier by making the corresponding substitutions and writing the new integrand in terms of s and t.

(c) Solve the equations $s = x + y$ and $t = x - y$ for x and y. (Doing so determines the standard form of the transformation, since we will have x as a function of s and t, and y as a function of s and t.)

(d) To actually execute this change of variables, we need to know the st-region D that corresponds to the xy-region D'.

  i. What st equation corresponds to the xy equation $x + y = 1$?
  ii. What st equation corresponds to the xy equation $x = 0$?
  iii. What st equation corresponds to the xy equation $y = 0$?
  iv. Sketch the st region D that corresponds to the xy domain D'.

(e) Make the change of variables indicated by $s = x+y$ and $t = x-y$ in the double integral (11.10) and set up an iterated integral in st variables whose value is the original given double integral. Finally, evaluate the iterated integral.

◁

Change of Variables in a Triple Integral

Given a function $f = f(x, y, z)$ over a region S' in $\mathbb{R}^3$, similar arguments can be used to show that a change of variables $x = x(s, t, u)$, $y = y(s, t, u)$, and $z = z(s, t, u)$ in a triple integral results in the equality

$$\iiint_{S'} f(x, y, z)\, dV = \iiint_{S} f(x(s, t, u), y(s, t, u), z(s, t, u)) \left| \frac{\partial(x, y, z)}{\partial(s, t, u)} \right| ds\, dt\, du,$$

where

$$\frac{\partial(x, y, z)}{\partial(s, t, u)} = \begin{vmatrix} \frac{\partial x}{\partial s} & \frac{\partial x}{\partial t} & \frac{\partial x}{\partial u} \\ \frac{\partial y}{\partial s} & \frac{\partial y}{\partial t} & \frac{\partial y}{\partial u} \\ \frac{\partial z}{\partial s} & \frac{\partial z}{\partial t} & \frac{\partial z}{\partial u} \end{vmatrix}.$$

In expanded form,

$$\frac{\partial(x, y, z)}{\partial(s, t, u)} = \frac{\partial x}{\partial s}\left[\frac{\partial y}{\partial t}\frac{\partial z}{\partial u} - \frac{\partial y}{\partial u}\frac{\partial z}{\partial t}\right] - \frac{\partial x}{\partial t}\left[\frac{\partial y}{\partial s}\frac{\partial z}{\partial u} - \frac{\partial y}{\partial u}\frac{\partial z}{\partial s}\right] + \frac{\partial x}{\partial u}\left[\frac{\partial y}{\partial s}\frac{\partial z}{\partial t} - \frac{\partial y}{\partial t}\frac{\partial z}{\partial s}\right].$$

The expression $\left|\frac{\partial(x,y,z)}{\partial(s,t,u)}\right|$ is again called the Jacobian.

Summary

- If an integral is described in terms of one set of variables, we may write that set of variables in terms of another set of the same number of variables. If the new variables are chosen appropriately, the transformed integral may be easier to evaluate.

- The Jacobian is a scalar function that relates the area or volume element in one coordinate system to the corresponding element in a new system determined by a change of variables.

Exercises

1. Let D' be the region in the xy plane that is the parallelogram with vertices $(3, 3)$, $(4, 5)$, $(5, 4)$, and $(6, 6)$.

    (a) Sketch and label the region D' in the xy plane.

    (b) Consider the integral $\iint_{D'} (x + y)\, dA$. Explain why this integral would be difficult to set up as an iterated integral.

    (c) Let a change of variables be given by $x = 2u + v$, $y = u + 2v$. Using substitution or elimination, solve this system of equations for u and v in terms of x and y.

(d) Use your work in (c) to find the pre-image, D, which lies in the uv plane, of the originally given region D', which lies in the xy plane. For instance, what uv point corresponds to $(3,3)$ in the xy plane?

(e) Use the change of variables in (c) and your other work to write a new iterated integral in u and v that is equivalent to the original xy integral $\iint_{D'} (x+y)\, dA$.

(f) Finally, evaluate the uv integral, and write a sentence to explain why the change of variables made the integration easier.

2. Consider the change of variables

$$x(\rho, \theta, \phi) = \rho \sin(\phi)\cos(\theta) \quad y(\rho, \theta, \phi) = \rho \sin(\phi)\sin(\theta) \quad z(\rho, \theta, \phi) = \rho \cos(\phi),$$

which is the transformation from spherical coordinates to rectangular coordinates. Determine the Jacobian of the transformation. How is the result connected to our earlier work with iterated integrals in spherical coordinates?

3. In this problem, our goal is to find the volume of the ellipsoid $\frac{x^2}{a^2} + \frac{y^2}{b^2} + \frac{z^2}{c^2} = 1$.

(a) Set up an iterated integral in rectangular coordinates whose value is the volume of the ellipsoid. Do so by using symmetry and taking 8 times the volume of the ellipsoid in the first octant where x, y, and z are all nonnegative.

(b) Explain why it makes sense to use the substitution $x = as$, $y = bt$, and $z = cu$ in order to make the region of integration simpler.

(c) Compute the Jacobian of the transformation given in (b).

(d) Execute the given change of variables and set up the corresponding new iterated integral in s, t, and u.

(e) Explain why this new integral is better, but is still difficult to evaluate. What additional change of variables would make the resulting integral easier to evaluate?

(f) Convert the integral from (d) to a new integral in spherical coordinates.

(g) Finally, evaluate the iterated integral in (f) and hence determine the volume of the ellipsoid.

Index

arclength, 81
average value over a solid, 238

center of mass
    of a solid, 238
change of variable
    double integral, 262
    triple integral, 264
Cobb-Douglas production function, 179
continuity, 102
coordinate planes, 6
critical point, 161
curvature, 86
cylindrical coordinates, 247

directional derivative, 149
discriminant, 163
dot product, 33
double integral
    difference in volumes, 186
double integral
    average value, 187
    center of mass of a lamina, 213
    definition, 186
    mass of lamina, 209
    probability, 215
double integral over a general region, 201
double Riemann sum, 184

function
    differentiable, 130
    domain, 3
    graph, 4
    locally linear, 130
    of two variables, 3
    surface, 4

graph of a vector-valued function
    definition, 62

iterated integral
    cylindrical coordinates, 249
    polar coordinates, 222
    rectangular coordinates, 192
    spherical coordinates, 253

Jacobian, 262, 264
joint probability density function, 214

Lagrange multiplier, 177
level curve, 12
line
    direction vector, 51
    in space, 51
    parametric equations, 54
    vector equation, 52
locally linear, 50

mass
    of a solid, 237
moments about coordinate axes, 213

parameterization
    curve, 62
    surface, 226
parametric equations for a curve, 62
partial derivatives

first-order, 108
second-order, 119
second-order, mixed, 119
second-order, unmixed, 119

plane
definition, 54
scalar equation, 55

polar coordinates, 219

projectile motion
parametric equations, 75

sphere
definition, 6
formula, 8

spherical coordinates, 250

surface area, 231

trace, 9

triple integral, 237
triple Riemann sum, 237

vector
angle between, 33
component in the direction of, 38
definition, 21
projection, 37, 38
subtraction, 24
sum, parallelogram, 25

vector-valued function
antiderivative, 73
definition, 62
derivative, 67
indefinite integral, 73

vectors
orthogonal, 35

volume of a solid, 238

Made in the USA
Lexington, KY
25 January 2018